T0138313

Shaping Science with Rhetoric

Shaping Science with Rhetoric

The Cases of Dobzhansky, Schrödinger, and Wilson

Leah Ceccarelli

The University of Chicago Press *Chicago and London*

Leah Ceccarelli is assistant professor of speech communication at the University of Washington.

The University of Chicago Press, Chicago 60637
The University of Chicago Press, Ltd., London

Printed in the United States of America

10 09 08 07 06 05 04 03 02 01 1 2 3 4 5

ISBN: 0-226-09906-7 (cloth)
ISBN: 0-226-09907-5 (paper)

Library of Congress Cataloging-in-Publication Data

Ceccarelli, Leah.
Shaping science with rhetoric : the cases of Dobzhansky, Schrödinger, and Wilson / Leah Ceccarelli.
p. cm.
Includes bibliographical references and index.
ISBN 0-226-09906-7 (cloth : alk. paper)—ISBN 0-226-09907-5 (pbk. : alk. paper)
1. Life sciences literature. 2. Rhetoric. 3. Interdisciplinary approach to knowledge. 4. Interdisciplinary research. 5. Wilson, Edward Osborne, 1929– Consilience. 6. Dobzhansky, Theodosius Grigorievich, 1900–1975 Genetics and the origin of species. 7. Schrödinger, Erwin, 1887–1961 What is life? I. Title.
[DNLM: 1. Interprofessional Relations. 2. Research—methods. 3. Communication. 4. Science—history. Q 180.55.I48 C387s 2001]
QH303.6 .C433 2001
507.2—dc21
00-012179

∞ The paper used in this publication meets the minimum requirements of the American National Standard for Information Sciences—Permanence of Paper for Printed Library Materials, ANSI Z39.48-1992.

Contents

Preface

This book is about interdisciplinarity—how scientists reach across disciplinary borders to motivate colleagues with very different intellectual and professional commitments to embark on new interdisciplinary lines of research. But interdisciplinarity is more than the subject of this book—it is its purpose as well. In this book I address scholars from my own area of research—rhetorical inquiry—as well as scientists, historians of science, and readers from various other areas of research who are interested in how new interdisciplinarity fields are formed.

This desire to speak across disciplinary borders may have something to do with my own history of interdisciplinarity. Years ago, when I was pursuing my bachelor's degree in cellular biology at the University of California at Berkeley, I took a course in the rhetoric department that set me on the path to writing this book. I was a bit skeptical about the course at first; could I learn anything of value from a department whose very name stands for the amoral "empty talk" that politicians use when they skirt the real issues in public debates? A friend who recommended the course told me to look up the word *rhetoric* in the dictionary before passing judgment. I found two definitions: (a) "mere bombast" and (b) "the study and/or use of effective speaking and writing." The department of rhetoric was named after the latter definition. In the first week of the course, I was shown that the discipline of rhetorical inquiry is part of an ancient tradition, with roots in Aristotle's *On Rhetoric,* a treatise that helps scholars develop "an ability, in each [particular] case, to see the available means of persuasion." It is a discipline that studies the intimate interconnection between words and thoughts; it seeks not to overcome truthful statements with tricks of language but to recognize the deep interdependence of our linguistic and nonlinguistic worlds.

Although the discipline typically studies speeches or essays from the public sphere, it occasionally turns its attention to written materials from the world of science. A few weeks into the course, we read and discussed Darwin's *Origin of Species.* As a biology student, I had encountered this touchstone of science before, but my experience this time was entirely different—we read the book not just to understand Darwin's evolutionary insights but to understand how he designed his arguments to persuade colleagues that his theory was correct. We discovered that Darwin was not merely a brilliant scientist, he was also an extraordinarily skilled rhetor, able to design arguments that built on the existing beliefs of his audience to persuade them to accept a new understanding of the natural world. After participating in a rhetorical study of Darwin's book, I was hooked, and though I have never lost my love for science, it now shares a space in my heart with my enthusiasm for rhetorical inquiry.

In this book, I combine my two passions to engage in the rhetorical study of three influential books by scientific writers. I draw from the tradition of *rhetorical inquiry* to explore how scientists, given the available means at their disposal, designed their arguments to persuade others. Though I look at a different type of scientific book than Darwin's *Origin*, I apply the same general categories of analysis that have been applied in rhetorical studies of that text. Recognizing that words and thoughts are integrally connected, my study of scientific writing traces the ways in which professional assumptions, linguistic resources, the constraints of the material world, cultural beliefs, and reasoned arguments are woven together by scientists when they seek to persuade others.

In writing this book, I have benefited greatly from the support and advice of many scholars from different disciplines. My work on Schrödinger began as a paper written for a graduate class by John Angus Campbell when he was serving as the Van Zelst professor at Northwestern University; he has continued to read my work and support my research long after that class, and I am very grateful for it. After Campbell's class, my work evolved into a master's thesis directed by Michael Hyde; then, with the addition of a second case study, it was expanded into a dissertation under the direction of Michael Leff. Thanks to the generosity of three granting agencies, a third case study was produced, and the work was transformed into a book manuscript. The first grant I received was a University of Washington Teaching Fellowship in the Center for the Humanities that allowed me to teach a course on the subject of interdisciplinarity. I benefited from insights generated by the graduate students who took that course: Kerry Godes, Amanda Graham, Jason Grant, David Lennox, and Susan McBurney. The second grant was a University of Washington Royalty Research Fund Scholar Award, and the third was a National Endowment for the Humanities Summer Stipend; both of these provided the much-needed time away from teaching that allowed me to concentrate on this project.

I would also like to thank Alan Gross and Steve Fuller, who published earlier versions of some of the work included in this book and provided commentary that helped me to see its strengths and attenuate its weaknesses. Parts of chapter 5 are revised from my article "A Masterpiece in a New Genre: The Rhetorical Negotiation of Two Audiences in Schrödinger's 'What Is Life?'" *Technical Communication Quarterly* 3 (winter 1994): 7–17, included here courtesy of the Association of Teachers of Technical Writing; an earlier version of chapter 3 was published as "A Rhetoric of Interdisciplinary Scientific Discourse: Textual Criticism of Dobzhansky's 'Genetics and the Origin of Species,'" *Social Epistemology* 9 (April–June 1995): 91–111, included here courtesy of Taylor & Francis, P.O. Box 25, Abingdon, Oxfordshire, OX14 3UE, United Kingdom, http://www.tandf.co.uk/journals. David Hull and John Beatty sent me in the right direction during my historical research on Dobzhansky's book, and Emma Baxter at Cambridge University Press went above and beyond the call of duty in ferreting through old files to uncover unpublished reviews of Schrödinger's book. Ken Alder, Charles Bazerman, Carol Berkenkotter, David Depew, and Carolyn Miller have all offered substantive critiques of my work that helped me improve my argument. Diane Barber answered a couple of questions about biology that were beyond my own bachelor's degree–level knowledge of the science, and Caspar Curjel translated some material from German to English for me. Others whose support directly contributed to the publication of this book include Jack Selzer, Keith Benson, and of course, my editor Susan Abrams, who patiently and adroitly walked me through the stressful process of getting my first book published. My greatest intellectual and professional debt is owed to my mentor, Michael Leff; any expression of my gratitude to him would be an understatement. Thanks also to my husband, Tim Smith, and to the many friends and colleagues who have provided the moral support I needed to reach my goal of completing this book and seeing it in print.

1

Inspiring Interdisciplinarity

With this book a new branch of scientific investigation came of age, and many workers were induced to enter it. The book is one of the few seminal publications of its generation.

Howard Levene, Lee Ehrman, and Rollin Richmond, "Theodosius Dobzhansky up to Now"

There is no other instance in the history of science in which a short semipopular book catalyzed the future development of a great field of research.

Walter Moore, *Schrödinger: Life and Thought*

Some books make an impact, not because they introduce new ideas that have never before been thought, but because they repackage ideas in a way that allows readers to see things they had previously been unable or unwilling to recognize. The two books referred to in the quotations above—Theodosius Dobzhansky's 1937 *Genetics and the Origin of Species* and Erwin Schrödinger's 1944 *What Is Life? The Physical Aspect of the Living Cell*—were such books. Both were addressed to groups of scientists from different fields who had conflicting intellectual and professional allegiances. Both were designed to help those scientists see beyond the barriers that separated their fields, urging them to change their professional goals and enter into new interdisciplinary alliances. And both books were highly successful at what they sought to do. Dobzhansky's book has been praised for motivating biologists from various fields to overcome their deep suspicions of each other and embark on "the evolutionary synthesis," an integration of research in genetics and natural history that resulted in the modern interdisciplinary domain of evolutionary biology. Schrödinger's book has been hailed for serving a similar function as the primary motivating influence that brought physicists and biologists together to do the research that would form the discipline of molecular biology.

Though both books were written by scientists, addressed to scientists, and concerned with science, neither performed the typical work of a scientific monograph. Neither author was primarily interested in introducing his unpublished research to a "core set" scientific community in the hope of establishing the truth of a novel scientific knowledge claim. Rather than monographs written to validate new scientific discoveries, they were monographs of broad interdisciplinary persuasion that sought to shape new scientific communities. Rather than make direct contributions to the history of ideas, these books attempted to change the way scientists perceived the nature and future of their own work and the work that colleagues were doing across disciplinary boundaries. Because the immediate response to each book was overwhelming support for the interdisciplinary research programs each promoted, these books are universally recognized by scientists and historians of science as two of the most influential works in twentieth-century science.

A more recent book with a similar purpose is Edward O. Wilson's 1998 *Consilience: The Unity of Knowledge.* Like Dobzhansky and Schrödinger, Wilson sought to build bridges between disciplines. Expanding on arguments that he made in the first and last chapters of his 1975 *Sociobiology: The New Synthesis,* Wilson built a case for uniting the social sciences and the humanities with evolutionary biology. But though he has had some success over the years with this argument, inspiring a number of readers to conduct research between fields, *Consilience* has not been an unqualified success at achieving its goals. Many readers of *Consilience* have rejected Wilson's call to action; they have been unpersuaded by his appeals for interdisciplinary collaboration. In fact, some have said they were so offended by *Consilience* that they have chosen to work actively against its goals. Whereas Dobzhansky and Schrödinger designed appeals that were accepted enthusiastically by their audiences, Wilson designed appeals that became the subject of intense controversy, accepted by some but firmly rejected by many others.

Given the inherently conservative bent of the academy when it comes to institutional structure and traditional departmental prerogatives, how do books such as Dobzhansky's and Schrödinger's stimulate interdisciplinary work without arousing suspicion and sparking controversy? Considering the investment most scientists have in the approaches, methods, and conceptual worldviews that define their own disciplines, what arguments can compel them to risk themselves, and their careers, in a research program that goes beyond traditional disciplinary boundaries? Why do some attempts to inspire interdisciplinary research achieve consensus of acclaim and support, while others are less successful at the same task, persuading only one group of readers while angering another?

In this book, case studies of these three authors' arguments for interdisciplinarity reveal what strategies work and do not work to persuade people to cross intellectual borders. These studies explain why books such as Dobzhansky's and Schrödinger's achieve such universal success with their audiences, while books such as Wilson's spark controversy. In addition to illuminating the rhetorical strategies best designed to motivate interdisciplinarity, these three case studies shed light on a previously unacknowledged and unexamined genre of scientific writing, one that, while failing to meet many commonly held expectations about scientific writing, has nevertheless had a significant impact on the history of science by helping to initiate new scientific communities.

Texts That Seek to Catalyze Community: An Unexamined Genre of Science

The rhetoric of science is a growing area of research, devoted mainly to the study of how scientists persuade each other.[1] Scholars in this field have done their best during the past two decades to prove the "hard case" that the prototypical scientific text, the research report that seeks to establish a scientific truth claim, is amenable to rhetorical scrutiny. On the whole, these scholars have done an excellent job of showing that rhetorical analysis *can* provide novel and intriguing commentary on the prototypical scientific text and its place in history.[2] Without adopting the view that science is *merely* rhetoric, most of those who read this literature recognize that scientists are indeed advocates for their own theories and data and that the more successful scientists are often those who can couch their findings in terms that are most persuasive to their peers. Some of the best research in the rhetoric of science undertakes the close reading of individual scientific texts to show exactly *how* they were designed to persuade specific audiences at particular moments in history to acknowledge the truth of their authors' theories.[3]

Unfortunately, in their rush to open the "hard case" of the prototypical scientific text to scrutiny, rhetoricians of science have failed to explore some of the other ways in which scientists use persuasion to advance the

1. For reviews of this literature, see Harris, "Rhetoric of Science"; A. Gross, *Rhetoric of Science*, reprint ed., vii–xxxiii; and Campbell and Benson, "Rhetorical Turn in Science Studies."

2. Several studies of Watson and Crick's "A Structure for Deoxy Ribose Nucleic Acid" show the variety of insights that can be made by different rhetoricians approaching a single prototypical scientific text. See Halloran, "Birth of Molecular Biology"; Bazerman, *Shaping Written Knowledge*, 18–55; Prelli, *Rhetoric of Science*, 236–57; A. Gross, *Rhetoric of Science*, 54–65; Miller, "*Kairos*"; and W. Fisher, "Narration."

3. John Angus Campbell's corpus of work on Darwin's *Origin* is a good example of scholarship that undertakes the close reading of a scientific text. For example, see Campbell, "Polemical Mr. Darwin"; Campbell, "On the Way." Greg Myers also provides some good examples of close reading in his *Writing Biology*.

scientific enterprise.[4] Ironically, the motivational texts scientists write to persuade their colleagues to undertake new research, texts that have an obvious rhetorical flavor, have been largely ignored by scholars of rhetoric.

Although historians of science have paid a bit more attention to this kind of text, they too have had difficulty identifying and assessing the genre. Historians acknowledge the motivational influence of books like Dobzhansky's and Schrödinger's, but because their discipline is not devoted to the analysis of subtle persuasive strategies, they cannot fully explain *how* such books achieved their influence. In attempting to account for the impact of such books, historians have often made broad pronouncements about the charismatic leadership of the author or the fortuitous timing of a book's publication. They have had little to say about the way in which the text itself was designed to achieve its influence.

Scientists also find themselves at a loss when it comes to explaining why these books achieved such influence. Though they recognize these books as motivational successes, they tend to assess all important works of science by the standards of the prototypical scientific truth claim. Using these standards of judgment, they struggle to explain why a work such as Schrödinger's, which fails to meet basic criteria of "good science" such as originality and accuracy, could have had such a strong and positive impact on science itself.

My study of books like these builds on a recognition that they belong to a special genre of scientific writing and should be understood and evaluated according to standards that are appropriate to that genre. We can call these texts *interdisciplinary inspirational* works of science, because books of this kind attempt to stimulate the growth of community between different scientific disciplines.[5]

A text in the interdisciplinary inspirational genre is like a catalyst—it addresses separate disciplines that are relatively inert and facilitates a reaction between them. Its main function is to encourage change, to motivate action in others. Like a catalyst, this type of text might seem at first glance to be a relatively minor part of science, but it can have a surprisingly large effect. When such a book succeeds, it produces a new area of research that otherwise would not have been formed or would have taken much longer to develop. The books by Dobzhansky and Schrödinger were designed to be incredibly effective catalysts, whereas the book by Wilson was written in a manner that undercuts its catalytic function.

In case studies of these three books, I identify several common characteristics of this newly identified genre, including a tendency toward

4. Charles Alan Taylor makes a similar point; see Taylor, "Science as Cultural Practice."

5. Scholarship in rhetorical inquiry suggests that genres are most adequately defined by the social action they seek to produce; that is why I have named the genre by its purpose. See Miller, "Genre as Social Action," 151.

synthesis rather than the introduction of original truth claims, the development of an authorial persona that is different from that used in the prototypical scientific text, and forms of address that recognize dual or even multiple audiences. This last characteristic is what leads to the most exciting finding from my analysis of this genre: two specific rhetorical strategies designed to appeal to divided audiences appear in the most effective texts but are absent in less effective texts.[6]

"Conceptual chiasmus" is a neologism I have constructed to describe the first of these, a rhetorical strategy that appears in the works of Dobzhansky and Schrödinger. A *chiasmus* is a rhetorical figure (a stylistic device) in which words and their order are reversed in parallel clauses: for example, "He went in, out went she." Each side of the sentence is a mirror image of the other. Biologists may be familiar with a similar term, *chiasma*, which, like *chiasmus*, is derived from the ancient Greek word for "cross." I propose "conceptual chiasmus" to indicate a rhetorical strategy that reverses disciplinary expectations surrounding conceptual categories, often through metaphor, to promote the parallel crisscrossing of intellectual space. With a conceptual chiasmus, unusual linguistic choices force readers from one discipline to think about an issue in terms more appropriate to their counterparts in another discipline, *and* vice versa. The result of this parallel reversal of disciplinary expectations is that the thought patterns of each side are forced temporarily to cross over to the other side. This crossing over may work to make interdisciplinary action more conceivable to readers from both fields.

The second rhetorical strategy I will identify, the polysemous textual construction, is similar in that it too works on different audiences differently. The word *polysemy* can be defined by breaking it into its roots: *poly*, meaning "many," and *semy*, having to do with meaning; *polysemy* is thus defined as "many meanings." A polysemous textual construction is a passage that can be read (that is, interpreted) in two or more ways.[7] When used effectively by an author, a polysemous passage can bring different audiences, for different reasons, to accept a message. Each group sees its own interests as being served by the polysemous passage; they interpret it differently (sometimes focusing on different parts of the passage to do so), and because of this, they unite in praise of the text and

6. The intentional fallacy keeps scholars from drawing conclusions one way or another about what an author consciously and deliberately did in a text. Recognizing this, rhetoricians often use the terms *strategy* and *design* without implying conscious intent on the part of authors. Just as an organism might adopt a successful evolutionary strategy without being consciously aware of it, so too might an author adopt a successful rhetorical strategy without being consciously aware of it. Dobzhansky and Schrödinger may or may not have deliberately inserted these rhetorical "strategies," which appeared in their texts and helped them achieve their goals.

7. For a more complete definition, see my article "Polysemy." In this book, I only discuss the kind of polysemy that I identify in that article as "strategic ambiguity."

are brought into a position of support that allows them to initiate an interdisciplinary alliance.

The action of both of these strategies—conceptual chiasmus and polysemous textual constructions—will become more clear in the case studies, where I closely analyze the rhetorical design of the texts and the responses of readers. As theoretical concepts, these rhetorical strategies may seem rather obscure, but the case studies will bring them to life, illuminating their presence and action in the structure of specific texts.

The Close Textual-Intertextual Analysis: Combining Rhetorical Criticism and Historical Research

Rhetoricians engage in a variety of approaches to analyzing texts. The "method" of analysis I adopt in the case studies to reveal how individual texts achieved their persuasive influence is a new approach to rhetorical criticism I call a "close textual-intertextual analysis." It is a variation on an established critical practice called "close textual analysis." The purpose of this modified approach is to explain how texts work by connecting rhetorical strategies to their effects on historical audiences.

Close Textual Analysis

Rhetoricians engaging in close textual analysis tend to focus on a single text at a time. By examining details of the text, they uncover subtle and otherwise unrecognized rhetorical strategies. They do this to explain how a text was constructed to invite a particular response in a particular audience. The close textual critic, in other words, scrutinizes a work in order to determine how its form was designed to achieve its function.

One way of understanding what a close textual analysis adds to the scholarly conversation is to think about it as the microscopic study of a primary text. Historians, social theorists, and cultural critics take an approach to scholarship that pieces together information from fragments of discourse across time. For example, historians of science are skilled at weaving a complex pattern out of the many traces of biographical, social, intellectual, and institutional influences on scientific thought. Uncovering multiple fragments of evidence, they synthesize explanations for scientific development that are expansive and intriguing. In contrast, close textual analysis offers the sustained examination of a single moment in history. Because close textual analysis restricts its focus to one work at a time, it can produce a microscopic study of that particular work that would not be possible if a broader range of materials were being discussed.

Of course, one problem with the critical approach of close textual analysis is that a researcher who spends so much time squinting through a microscope tends to become somewhat myopic. The critic who focuses too much attention on the internal characteristics of a particular work often neglects to fully explore the external influences that the text had on its context, or the external influences its context had over it.[8] Marveling at the formal intricacy of a textual strategy that appears to be perfectly designed, a rhetorician may be tempted to proclaim its author a master rhetor and to judge the text an exemplary product. But it can be a mistake to make this sort of judgment before examining how the audience actually interpreted the work in question. The close textual critic can only say how an audience was *invited* to respond; the critic is unable to make any conclusions about the actual persuasive influence of the text. For various reasons, some having to do with the design of the text and some not, an audience may have chosen to decline the invitation.[9]

This limitation with close textual analysis is even more problematic when a text has multiple audiences. It may be the case that an ideal reader identified by the author was invited to interpret a text in a particular way, but the text was received very differently by other groups of readers with different interests and backgrounds.[10] A close textual analysis that examines a text from the perspective of the reader who was conspicuously constructed in that text may neglect to recognize the way in which other audiences were invited (deliberately or not) to interpret the text in different ways. The influence of a text on the world (and the influence of the world on a text) is often more complicated than the close textual critic imagines when closely scrutinizing the inner patterns of the text.

In short, the close textual critic can postulate how the intrinsic design of a text (its form) is connected to its extrinsic effect (its function). But without looking beyond the text itself, a close textual critic can make no confident claims about that connection. Also, without looking at the reception given a work by its actual heterogeneous audiences, one can come away from a close textual reading with an understanding of its compositional artistry that takes an unnecessarily limited view of what it means and what it does in the world.

Adding the Study of Reception to Close Textual Analysis

Close textual critics who seek to escape the decline into a myopic formalism will often supplement the close reading of a text with a study of context. But rhetoricians usually conduct this research by reading the work of historians, relying on secondary sources for the relevant information

8. Michael Leff, a practitioner of close textual analysis, puts it well when he acknowledges that the approach unwittingly sponsors "a local formalism that isolates the text from larger discursive formations and restricts interpretation within the orbit of the text's own construction." Leff, "Things Made by Words," 228.

9. The rhetorician of science Alan Gross recently voiced a concern about this approach to studying scientific texts when he warned that "the careful unraveling of the verbal microstructure of scientific texts, whatever it tells us about the intent of authors, can say little about its effect on readers." Gross, *Rhetoric of Science*, reprint ed., xvii.

10. Condit, "Rhetorical Criticism," 334–35.

about the situation in which the text was produced and read. From this research, critics learn important things about the interests and attitudes of people in the culture surrounding the text. However, rhetoricians rarely take this research further; few have uncovered intertextual material produced in response to a primary text, and fewer still have conducted a close reading of that material. This is where I recommend a modification to the "close textual analysis" approach to criticism. I urge rhetorical critics to explore all available evidence of the reception to a work; we should conduct a close textual analysis not only of the primary text, but also of the intertextual material produced by audience members who were responding to it.

This new approach to rhetorical criticism, a "close textual-intertextual analysis," provides a more reliable connection between internal form and external function. The rhetorician conducts a close reading of the text in its context to offer hypotheses about how readers might have been invited to respond to the text's appeal. The rhetorician then tests these hypotheses through a close reading of contemporary responses, such as book reviews, speeches, editorials, articles, or letters that make direct reference to the primary work under examination.[11]

There are at least two convenient forms of intertextual material that preserve the response of audiences to the scientific books I examine in the case studies: scientific articles that cited the primary texts, and book reviews. I consulted the Science Citation Index and uncovered more than sixty articles that cited these three texts in the years immediately following their publication. I also uncovered more than eighty book reviews that recorded reader response to the three primary texts. My intent was not merely to *show* that these monographs have been extensively cited and reviewed by scientists, but to discover *how* readers interpreted each book and to identify the aspects of each book that produced specific responses. I checked each hypothesis about how a particular rhetorical strategy invited a particular response against evidence of the actual response; at the same time, evidence of each response allowed me to reexamine the text itself to look for strategies that I otherwise might have failed to fully recognize.[12]

11. This critical practice of moving between text and response is not an entirely new invention. Literary critics familiar with reader-response criticism and reception studies will recognize the terms of this solution to the defects of formalist approaches. My approach shares with these approaches the desire to locate meaning in the interpretations of audiences, rather than harden meaning in text, author, or critic. My approach also has affinities with the new audience research in scholarship on popular culture texts, which seeks to uncover and validate the experiences of real audiences. For more on the similarities and differences between these approaches and mine, see my concluding remarks on critical practice in chapter 9.

12. Although I closely read more than sixty scientific articles that cited the books in question, I only found a third of them to be useful to the case studies. This was not unexpected. It is rare for a prototypical scientific article to include personal information, and so it is rare to find direct evidence there of influences on something as personal as a choice to

Because this approach to scholarship takes the rhetorician's interest in analyzing the persuasive strategies of influential primary texts and adds the historian's concern with tracing evidence of influence in the historical record, it benefits from best of two worlds: it supplements studies in the history of science with a rhetorical microanalysis, while supplementing rhetorical studies with the historian's attention to documentary detail.

Each case study is centered on one primary text and is divided into two chapters. The first of the pair includes a historical analysis of the primary text and its context, and the second is a close rhetorical reading of the text and its reception. Although each case study can be read on its own, the third one draws heavily on the first two, and is therefore best understood if read in sequence. When all three case studies are brought together, they support broader claims about this genre of scientific text and how it inspires interdisciplinary activity among scientists.

---

engage in interdisciplinary research. However, I did find that in many of the scientific articles that cited the primary texts, I could draw conclusions about the way in which their authors understood the meaning of the texts, and I could note what aspects of the texts most attracted their attention. On the whole, book reviews were more useful for determining what readers thought about a text, because the genre is designed to explicitly interpret and judge a text. For that reason, more book reviews are used in the case studies than scientific articles that cited the texts, but both are used to some degree for the intertextual component of these case studies. Later claims of influence in essays or autobiographies or oral narratives by scientists who were reflecting on their earlier career choices are used in the case studies as well. But since memories may prove less accurate than more contemporaneous evidence of the effect of these texts, I use such evidence with caution. In the case studies, remembered influences are used more often to establish the fact that people think these texts served a motivational function, and less often to test hypotheses about how particular rhetorical strategies worked on readers.

I Theodosius Dobzhansky's *Genetics and the Origin of Species*

2

The Initiator of the Evolutionary Synthesis: Historians and Scientists Weigh In

With the benefit of hindsight, it is easy to imagine a simple order in the history of science and to tell an equally simple story. For example, the following story is the one I heard as an undergraduate biology major being educated in the fundamentals of my chosen field:

Charles Darwin wrote *On the Origin of Species* to set out his discovery of evolution. Initially, there were battles over the religious heresy of his claim that it was natural selection, not divine intent, that was the driving force behind the design of biological life. But educated people soon accepted his theory and evolution became entrenched as scientific knowledge. Of course, the scientists of Darwin's time knew little about the hereditary mechanism itself; they knew only that natural selection worked on *something* to change a species over time. When scientists rediscovered Mendel's research on peas, the field of genetics was formed, and it provided the explanation of heredity that Darwin lacked. From genetics, we learned that mutant genes were the material used by natural selection to shape a new species. The rough edges of evolutionary theory were smoothed over, and the way was paved for molecular biology to take the next step in the forward march of knowledge.[1]

The story is not wholly wrong, but in describing events as if they fell into place neatly and without conflict, it provides a simplistic treatment of the people and ideas that led to the twentieth-century scientific acceptance of a unified evolutionary theory. Looking at our current understanding of scientific knowledge and reading out a story that neatly describes the history of science as a straight path toward truth is not the same as studying the opinions and commitments of historical figures

1. William B. Provine points out that this story is often found in textbooks. Provine, "Epilogue," 405.

and acknowledging the many twists and turns they took along the way to eventual agreement. Scholars who study this period of history point out that there was no long and complacent Darwinian tradition that was capped off in the early twentieth century with the weight of accumulated knowledge.[2] In the first place, the history of evolutionary biology is not over, and probably never will be. Though the latter half of the twentieth century may have seen a fairly stable agreement about evolutionary mechanisms, that agreement does not promise to remain indefinitely steady.[3] In addition, it is clear that though the idea of evolution was accepted by most educated people soon after Darwin's book was published, the theory of natural selection was not universally accepted during Darwin's time or even after Mendel's laws were rediscovered.[4] Instead, for about eighty years there was a great deal of conflict and dissent among people who worked in different disciplines within the life sciences and held allegiances to different theories of evolution. As this chapter reveals, our modern understanding of evolution was not built smoothly with each new discovery but came together all at once during a relatively short span of time in the 1930s and 1940s. What is especially remarkable about this "evolutionary synthesis" (sometimes called the "modern synthesis") is that it was so largely inspired by a single book that offered no new discovery or theory of its own.

In the interest of telling a more nuanced story than the one offered in an introductory biology class, this chapter moves from a broad and unfocused picture of the evolutionary synthesis to a more fine and focused image of the ideas and events that caused it. It begins with a narrative about the development of modern evolutionary theory, one that is only somewhat more detailed that the one above.[5] The chapter then moves to examine the unresolved historical debate about the general cause of that development. It refines the focus a bit further to see how scholars discuss a more proximate cause, Theodosius Dobzhansky's influential *Genetics and the Origin of Species.* Finally, this chapter suggests that a further tightening of the focus is necessary to fully understand the way in which Dobzhansky's text influenced the history of science. That tight focus is drawn in

2. For example, see Gould, introduction to *Genetics and the Origin of Species,* xix.

3. Vassiliki Betty Smocovitis describes the explosive appearance in the 1980s of "anti-synthesis" literature; see Smocovitis, *Unifying Biology,* 33–39. Richard M. Burian reviews some of this literature; see Burian, "Challenges."

4. J. Moore, "History of Evolutionary Concepts," 11.

5. As Debra Journet points out, "the history of the synthesis is itself a rhetorical construct. . . . No history is neutral, and histories of the synthesis are often constructed in order to establish priority or justify current approaches." Journet, "Synthesizing Disciplinary Narratives," 117. With this in mind, I have tried not to rely solely on secondary sources but to also turn to primary sources whenever possible. I have also tried to weave together a number of different accounts, from people with different interests. My narrative is plausible and fits most closely with the historical documents I have encountered, but I would never claim that it is the only story that could be told.

the next chapter's close rhetorical reading of the text and its intertextual antecedents and responses.

Conflict between Disciplines and Theories

> On the whole—and admittedly this is an oversimplification—two camps were recognizable, the geneticists and the naturalists-systematists. They spoke different languages; their attempts in joint meetings to come to an agreement were unsuccessful. In the early 1930s, despite all that had been learned in the preceding seventy years, the level of disagreement among the different camps of biology seemed almost as great as in Darwin's days. And yet, within the short span of twelve years (1936–1947), the disagreements were almost suddenly cleared away and a seemingly new theory of evolution was synthesized from the valid components of the previously feuding theories.[6]

As Ernst Mayr and William B. Provine explain in their historical account, conflict existed between at least two groups of biologists in the years preceding the evolutionary synthesis. The historian Garland Allen confirms that at the turn of the century and beyond, a wide chasm separated the more descriptive naturalists and the more experimental geneticists.[7] The differences between the two camps made it difficult for each side to understand and effectively refute the arguments of the other.[8] Studies of this period have shown that this breakdown of communication was driven by both methodological disagreements and ideological commitments.

With regard to scientific practice, the research traditions of the two groups were diametrically opposed. While the naturalist studied evolution through field and museum observations of natural populations, the geneticist studied heredity through experimental manipulations of laboratory populations.[9] Because geneticists were committed to the newer, more positivistic method, they had a stronger position in the fight over resources. Their experimental techniques were favored by funding authorities, such as the Rockefeller Foundation, and their ability to produce quick results attracted students.[10] At academic institutions in America, experimental

6. Mayr and Provine, preface to *Evolutionary Synthesis,* ix. One way in which this is an oversimplification is their decision not to mention other scientists who interacted with geneticists and naturalists during this period. The biometricians and the embryologists were both important institutional communities that were distinct from the geneticist and naturalist traditions, and both were eventually integrated into modern evolutionary theory. A study of the interdisciplinary discourse that marked the transitions in these fields, though much needed, is beyond the scope of this book. For a start along this line of research, see Sapp, "Struggle for Authority," 311–42.

7. Allen, "Naturalists and Experimentalists," 179; Allen, "Theodosius Dobzhansky," 87–98.

8. Mayr, *Growth of Biological Thought,* 541.

9. Ibid., 542; Allen, "Naturalists and Experimentalists," 185.

10. Abir-Am, "Discourse of Physical Power," 350; Sapp, "Struggle for Authority," 335.

sciences were increasingly favored over natural history and other descriptive sciences, so that by the 1920s and 1930s, naturalists who studied evolution were faced with the possibility of losing their position in the university.[11] With the same scarce resources being sought by each group, the battle over academic territory became fierce.

Allen does a thorough job of documenting the terms of this battle from the turn of the century to the 1920s.[12] He points to William Bateson's 1914 characterization of naturalists as "expounding teleological systems which would have delighted Dr. Pangloss himself" as an example of the scorn that geneticists held for their counterparts.[13] Also typical of the divide between the two groups was the paleontologist William K. Gregory's response to Bateson, defending the work of naturalists as a distinct and important line of investigation that "will gradually reassert [itself] even in competition" with geneticists.[14]

During the synthesis years (1936–1947), both naturalists and geneticists who read Dobzhansky's book had these battles on their minds. In 1944, the paleontologist George Gaylord Simpson described the vast divide between the methods of naturalists and experimentalists, a division that he recalled being most severe during the 1920s. According to Simpson, paleontologists had believed that geneticists were working on a pursuit too far removed from nature: "a geneticist was a person who shut himself in a room, pulled down the shades, watched small flies disporting themselves in milk bottles, and thought that he was studying nature." On the other hand, Simpson said, geneticists had believed that paleontologists were involved in a futile attempt to study a complex system through nonscientific description of external appearances: "like a man who undertakes to study the principles of the internal combustion engine by standing on a street corner and watching the motor cars whiz by."[15] In 1941, Leon J. Cole, a geneticist, described his own experience with the separation between museum and laboratory workers, or "species splitters" and "sperm chasers," as they had called each other:

> The field naturalist and museum worker felt that the "closet zoologist," as he dubbed the laboratory investigator, was working under such artificial conditions that his findings had little relation to animals in "natural" surroundings. Above all, that he had little conception of taxonomic problems, and the chances were that if he should meet in the wild the animal on whose tissues he was working he would not recognize it.

11. Smocovitis, "Unifying Biology," 14; Sapp, "Struggle for Authority," 340.

12. Allen, "Naturalists and Experimentalists," 181–86; see also Sapp, "Struggle for Authority," 327–34.

13. Bateson, "Address of the President," 293.

14. W. K. Gregory, "Genetics versus Paleontology," 622–23.

15. Simpson, *Tempo and Mode*, xv.

On the other hand, the geneticist and other laboratory workers

> accused the taxonomist of playing a game about on a par with arranging a collection of postage stamps, and of splitting genera into species, and species into subspecies, merely for the doubtful distinction of getting his name attached to the specimens and embalmed for posterity and for eternity along with them.[16]

As these comments indicate, each camp considered its own methods to be superior to the other's. The "evangelical" laboratory workers defined themselves in opposition to the older, "nonprogressive" field and museum workers who merely used "descriptive" and "observational" techniques.[17] There was no doubt in the minds of the experimental biologists that their more objective methods were superior to the "speculative" approach of the evolutionary naturalist.[18] In contrast, the naturalists saw the isolation of the laboratory worker as a barrier to understanding the complexity of nature. Since laboratory workers were studying a problem out of its original context, naturalists believed that the generalizations they derived were overly restricted and hopelessly oversimplified.[19]

A study of the history of ideas during this period shows that the differences in scientific approach between these two groups paralleled a difference in commitment to particular evolutionary theories. Many naturalists were Darwinian, supporting the theory that natural selection worked on the small continuous variations in characteristics that were observed in natural populations. They thought that the large-scale mutations studied by geneticists in the laboratory were "monstrosities" and "sports" that would quickly die in the real world. In contrast, geneticists held allegiance to theories that were supported by Mendelian laws, arguing that the slight, continuous variations observed in natural populations were not inherited and thus played no role in the evolutionary process. In early years, they even tended to support de Vries's theory of evolution by macromutation, in which large-scale jumps, or "saltations," produced new species in a single generation.[20] According to Allen,

> This rigid and artificial distinction between continuous and discontinuous variation, which so many workers insisted on making in this period, was a serious hindrance to any integration of Mendelian and Darwinian theory. So long as selection was seen as acting on either one or the other type of variation, the Mendelian concept

16. Cole, "Each after His Kind," 290.
17. Cain, "Common Problems," 16.
18. Mayr, "Prologue," 10; Beltrán, "Naturalist in America," 547; Smocovitis, "Unifying Biology," 16.
19. Allen, "Naturalists and Experimentalists," 182.
20. Ibid., 180; Provine, "Role of Mathematical Population Geneticists," 172; Mayr, "Role of Systematics," 124; J. Moore, "History of Evolutionary Concepts," 13.

(representing discontinuity) and the Darwinian concept (representing continuity) could only be seen as opposite and mutually exclusive theories.[21]

And as long as geneticists exclusively studied discontinuous mutation in laboratory populations and naturalists exclusively observed continuous variation in natural populations, the divide was likely to continue.

Another conceptual commitment that deepened the chasm between the two camps was likewise based on differences in disciplinary approach. The research traditions of the two groups led them to focus on different dimensions of evolution and to study different levels in the hierarchy of natural phenomena.[22] To recognize this distinction, we must first understand that naturalists and geneticists asked different questions of their biological subject matter. According to Mayr, naturalists were interested in "ultimate causation," asking historical questions about why nature was organized as it was. Geneticists were interested in "proximate causation," focusing on the functional question of how things worked.[23] The result of this difference in emphasis was the formation of both temporal and spatial rifts between the two groups.

In the temporal dimension, the naturalists' historical questions led to a diachronic focus on phylogeny. They studied evolutionary relationships through time, explaining common descent with the construction of family trees that outlined the manner in which species shared ancestors.[24] Geneticists were preoccupied with a synchronic focus on the mechanism of heredity. They took a cross-section of evolutionary time, focusing on the apparatus at work during a single moment in a single species.[25] According to the historian Jan Sapp, the distinction between "phenotype" and "genotype" was originally invented by a geneticist to promote his own mechanistic study of heredity and denigrate the historical view of the naturalists.[26] What began as a polemical distinction was eventually adopted as a commonsense description of the differences between the two fields; naturalists studied the historical evolution of the phenotype (the characteristics of organisms), geneticists studied the genotype (pool of genes) to uncover the fundamental mechanisms of heredity.[27]

At the spatial level, this distinction could also be used to show how the two groups focused on different parts of the hierarchy that organizes biological matter. While naturalists studied species and taxa, geneticists studied the "pool of genes" within an organism. Naturalists were more

21. Allen, "Naturalists and Experimentalists," 187.

22. Mayr, *Growth of Biological Thought*, 542.

23. Mayr, "Prologue," 9; see also Allen, "Theodosius Dobzhansky," 90.

24. Allen, "Naturalists and Experimentalists," 134; J. Moore, "History of Evolutionary Concepts," 14.

25. Mayr, "Prologue," 13.

26. Sapp, "Struggle for Authority," 322–23.

27. Allen, "Naturalists and Experimentalists," 201.

prone to study large groups and more likely to recognize the variability that exists in a population. However, they did not recognize the underlying mechanisms that produced that variation. Geneticists, by virtue of their laboratory orientation and attention to the genotype, were more likely to study small groups and to experiment with "pure lines" of identical organisms. As a result, they understood hereditary mechanisms but often failed to recognize the variation that exists in natural populations.[28]

With these strong differences in methodology, in form of evolutionary theory favored, and in dimension of study, the naturalists and the geneticists were unlikely to combine their findings to bring the study of evolution into any coherent order. By 1930, the gap was so deep and wide that it looked as if nothing would be able to bridge it. But during the late 1930s and the 1940s, the chasm disappeared and a consensus over the mechanisms of evolution emerged.[29] Misunderstandings were removed, bridges between hierarchical levels were built, the warring camps were reconciled, and the "evolutionary synthesis" was constructed.

The Evolutionary Synthesis

The evolutionary synthesis has been called one of the most important events in the history of twentieth-century biology.[30] The central point of the synthesis was that Mendel's genetics and Darwin's theory of evolution by natural selection could be joined to explain all evolutionary phenomena.[31] As a result of the synthesis, two conclusions were generally accepted by both camps: that gradual evolution could be explained in terms of small mutations and the recombination of chromosomes, and that higher-level evolutionary phenomena such as the formation of new species could be explained in a matter consistent with known genetic mechanisms.[32] It convinced the geneticists to abandon their belief that evolutionary change was discontinuous and therefore unlike the variations studied by naturalists, and it convinced the naturalists to abandon their belief that laboratory experiments on mutations were irrelevant to the sorts of variation and evolutionary change found in natural populations.[33]

To understand the nature of the evolutionary synthesis it is crucial to recognize that it did not signal the triumph of one field over the other, nor was it immediately sparked by a new theory or discovery. Instead, the synthesis was an interdisciplinary agreement that cooperation between paradigms was both possible and desirable.

Some scholars have read the evolutionary synthesis as the triumph of

28. Ibid.; Mayr, "Role of Systematics," 128.
29. Provine, "Francis B. Sumner," 212.
30. Provine, "Epilogue," 399–400; Provine, "Francis B. Sumner," 212.
31. Burian, "Influence," 155.
32. Mayr, "Prologue," 1.
33. Burian, "Influence," 156.

genetics over the subject matter of natural history. But the philosopher Dudley Shapere argues that this is a misreading of the event:

> philosophers (and sometimes scientists) have limited themselves by thinking that the only kinds of relations of unification between theories are relations of reduction, where an old theory is absorbed into (its concepts defined in terms of, and its postulates deduced from) a newer or more general one. In the evolutionary synthesis, however, there was a mutual modification or supplementation of different theories (and, more generally, of the concepts, techniques, problems, and so forth, of different areas).[34]

Natural historians were the first to argue this case, suggesting that the standard "victory" story supports the interests of the more institutionally powerful laboratory sciences but ignores the contributions of the less powerful disciplines. For example, Mayr argues that the synthesis was not the conquest of one paradigm by another but a fair "exchange of the most viable components of the previously competing research traditions."[35] Stephen Jay Gould describes the conventional impulse among scholars to apply a reductionist bias to the history of science and proclaim that the more institutionally powerful quantitative and experimental disciplines have consistently conquered the more holistic and historical fields: "In this context, genetics is superior to natural history—and the modern synthesis must represent an imposition of genetic truths upon a static and musty (if not downright benighted) group of taxonomists. Thus, the modern synthesis is often portrayed as a unilinear transfer of truth, an irresistible genetic proclamation of Darwinism."[36] But this story is misleading, if not downright false, says Gould. "The synthesis was a true fusion of genetics and natural history, not an imposition of one progressive field upon another hidebound profession."[37]

Although these arguments are made by the alleged "losers" in the power struggle, I think they serve as a reasonable corrective to the strict reductionist account. To achieve the synthesis, both sides had to give up something. The most accurate understanding of the synthesis seems to lie somewhere between the two extremes of total triumph and true fusion. One group clearly attained and maintained a position of institutional power; but the end result of the conflict was nothing like a total surrender. The synthesis was somewhat uneven, but it was a synthesis nonetheless.

Another misleading impulse is to treat any significant event in the history of science as a paradigm change sparked by the discovery of a new fact or theory. But in the case of the evolutionary synthesis, there was no missing piece of information that needed to be found, nor was there a

34. Shapere, "Meaning," 396.
35. Mayr, "Prologue," 40.
36. Gould, introduction to *Genetics and the Origin of Species*, xix.
37. Ibid., xxxiv.

neat, elegant, critical experiment that settled the controversy.[38] Scholars are hesitant to even name this event the "synthetic theory of evolution" or the "synthetic paradigm" since there was no single, well-articulated new theory or disciplinary matrix that came out of this event.[39] David Depew and Bruce Weber effectively disclose the nature of the synthesis when they name it a "treaty" that allowed naturalists and geneticists to work together under a common set of interdisciplinary presuppositions.[40] Similarly, William Bechtel effectively describes the product of the synthesis when he speaks of the resulting "interdisciplinary research cluster."[41]

Rather than a scientific theory that required the discovery of new knowledge or the acceptance of a new scientific fact, the "evolutionary synthesis" was a movement that reorganized disciplines, overcoming intellectual and professional barriers that were keeping scientists in different areas from working together. It was a conceptual and political understanding that resulted in collaboration between disciplines, rather than a novel truth claim authorized by an invisible college of workers within a particular area.

What Launched the Synthesis?

In the secondary literature surrounding the evolutionary synthesis, the struggle to define it is surpassed only by the struggle to explain what was responsible for bringing it about. Historians, scientists, and philosophers have proposed several explanations for this unexpected interdisciplinary agreement; some say that developments in theory were responsible for sparking the synthesis, others point to experimental and field discoveries that offered new information to the scientific community, still others focus on conceptual changes that cleared the way for collaboration, and a final group recognizes social forces as an important influence on those who would form the synthesis.

Although scholars do not define the evolutionary synthesis as a "theory," some argue that developments in theory were responsible for the interdisciplinary treaty. The theory most often cited is the work of the mathematical population geneticists J. B. S. Haldane, R. A. Fisher, and Sewall Wright.[42] By designing mathematical models of natural populations, these geneticists were able to show that natural selection and mutation were different parts of the same system. With a formula called the "Hardy-Weinberg Equilibrium," they quantified parameters that were

38. Mayr, "Prologue," 39; J. Moore, "History of Evolutionary Concepts," 7.

39. Bechtel, "Editor's Commentary," 137; Shapere, "Meaning," 392–94; Burian, "Influence," 149.

40. Depew and Weber, "Consequences," 317. See also Depew and Weber, *Darwinism Evolving*, 300; Burian, "Challenges," 248; and Burian, "Influence," 156.

41. Bechtel, "Editor's Commentary," 137.

42. For example, see Provine, "Role of Mathematical Population Geneticists," 167–92; Allen, *Life Science*, 134–43.

usually separated by the two camps and demonstrated that theoretical concepts once believed contradictory could actually reinforce one another.[43] Over time, this formula was acknowledged by biologists in different disciplines, and the evolutionary synthesis was formed.

Another advancement often held to be responsible for the evolutionary synthesis was a series of field discoveries made by the ecological geneticist Sergei Chetverikov and others. His school of population genetics fruitfully combined the methods of geneticists and naturalists by not only developing mathematical models but testing those models through field observations. Because his Russian publications did not always reach biologists working in the West, his influence was not as strong as it might have been, but his field data were said to have laid the foundations for the synthesis.[44]

Data from other fields have also been cited as the cause of the evolutionary synthesis. The philosopher Lindley Darden makes the argument that the synthesis was not formed as the result of any single new theory but because the development of knowledge in many different disciplines had reached a significantly advanced stage. The synthesis coalesced once scientists could look to "Mendelian genetics for mutations; cytology for chromosomal abnormalities; [and] mathematical population genetics and experimental and field studies of populations for the populational level" of evolutionary phenomena.[45] Because each discipline had advanced to an appropriate stage of development, the information existed to form a synthesis of evolutionary theory.[46]

Each of these historical accounts marks some development in theoretical model or disciplinary information as the driving cause of the synthesis. Another hypothesis is that the synthesis required a broader conceptual change on the part of scientists working in the relevant disciplines. For example, Garland Allen argues that before the beginning of the synthesis, many biologists were applying a mechanistic bias to their treatments of complex populations. In the attempt to make biology more like classical physics, geneticists treated populations as uniform types and imagined genes to be atomistic units, similar to molecules of gas in a container. What was required to develop an advanced understanding of evolution was a rejection of this simplistic language and a recognition of the complex dynamics that exist both within a population and between genes. Scientists had to stop imagining groups of atomistic genes and uniform individuals so that they could develop a holistic understanding of evolution in terms of interacting genes and complex populations in their environments.[47]

43. Smocovitis, "Unifying Biology," 23.
44. Allen, *Life Science*, 129–34; Mayr, *Growth of Biological Thought*, 556–59.
45. Darden, "Relations among Fields," 115.
46. Ibid., 121.
47. Allen, *Life Science*, 144–45.

The historian Betty Smocovitis suggests that the move needed to bring about an evolutionary synthesis was actually a conceptual leap in the opposite direction. According to her, the biological disciplines were fractured into two camps because natural history was lagging behind in the twentieth-century struggle to make biology a "harder" science. What was needed to bring the two camps onto the same level was a capacity on the part of scientists to think of evolutionary theory in a manner that more closely approximated physics. To accomplish this, the language of physics was applied to evolutionary concepts and the Hardy-Weinberg formula was used to produce Newtonian-like laws of evolution. For example, natural selection was mechanized with words such as *cause*, *force*, and *mechanism* to purge it of the metaphysical taint of agency imposed on it in Darwin's time. The "force" of natural selection was then measured in equilibrium equations. This purification of suspect language in the evolutionary fields redeemed biology as a science and allowed the warring camps to unite under the umbrella of an evolutionary synthesis.[48]

The apparent contradiction between the accounts of Allen and Smocovitis regarding the value of mechanistic thinking in the production of the evolutionary synthesis need not overly concern us. A pluralist reading would argue that the geneticists needed a less mechanistic mode of thinking to appreciate the dynamic complexity of populations, and the naturalists needed a more mechanistic understanding of natural history to unite with the more prestigious experimentalist sciences.

A final explanation for the formation of the evolutionary synthesis moves away from the study of theoretical, informational, or conceptual developments to offer a study of social factors. For example, the historian Joseph Allen Cain suggests that institutional cooperation was spurred by sociopsychological motives. With the war between camps becoming increasingly fierce, the museum and field workers began to lose their status and their resources to the upstart laboratory sciences. In no time, the experimentalists seemed to be in secure control of biology's mainstream. Natural historians, unwilling to become second-class researchers, sought to redefine that mainstream and preserve their status with a new "synthesis" of evolutionary science. "In other words, for some the drive for cooperation and synthesis was (partially) motivated by a desire to ensure their inclusion within the evolutionary studies community. Collaboration offered a route to equal status."[49]

Although some scholars argue that science develops strictly on a cognitive level, and others argue that scientific change is wholly political, I believe that explanations on the level of conceptual development and explanations on the level of social motive can exist together in a coherent treatment of the history of science. Like other events in science, the inte-

48. Smocovitis, "Unifying Biology," 20–27.
49. Cain, "Common Problems," 18–19.

gration of disciplines that occurred during the synthesis was both conceptual and social in nature.[50] The fact that scholars offer explanations on different levels might be taken as evidence that there was no single cause of this historical event. A variety of factors—theoretical, informational, conceptual, and social—were needed to initiate the evolutionary synthesis.

We begin to see the first outlines of a connection between these different explanations when we recognize that there is one book that everyone mentions when talking about the evolutionary synthesis. Whatever their hypotheses regarding the underlying cause of this event, scholars who write about the history of science always cite a single document as the definitive inception of the evolutionary synthesis. This publication was an evolutionary treatise that did more to influence interdisciplinary agreement than any other; it got the word out about developments in theory and data in the different disciplines, built a new conceptual understanding in the minds of its readers, and convinced people from different social groups that it was in their best interest to draw disciplines together. This book was Theodosius Dobzhansky's *Genetics and the Origin of Species.*

The Influence of Dobzhansky's *Genetics and the Origin of Species*

According to Provine, the most influential book on evolution in the twentieth century was Dobzhansky's *Genetics and the Origin of Species:*[51] "[It] was, at least in the United States, the single most influential book of the evolutionary synthesis. The book was required reading for all evolutionists (I have yet to find an evolutionist trained in the United States between 1937 and 1960 who did not read the book), and it received widespread acclaim."[52] Earlier texts had treated the subject of population genetics and evolutionary theory—texts such as R. A. Fisher's *Genetical Theory of Natural Selection,* Sewall Wright's 1931 and 1932 papers on population genetics, and J. B. S. Haldane's 1932 *Causes of Evolution*—but none had the impact of Dobzhansky's book.[53] Dobzhansky's treatise has been called "one of the few seminal publications of its generation," a "milestone of progress in our understanding of evolution," and a "legacy" that will persist long after its author and original readers are gone.[54] It has been compared in importance to the nineteenth-century treatise from which it drew its name—Charles Darwin's *Origin of Species.*[55] But perhaps the most telling comment on the tremendous influence of this book was heard at a 1974 conference on the evolutionary synthesis, where biologists who

50. Bechtel, "Editor's Commentary," 137.

51. Provine, "Role of Mathematical Population Geneticists," 180.

52. Provine, "Origins of the Genetics of Natural Populations Series," 74.

53. Glass, introduction to *The Roving Naturalist,* 1; Ayala, "Nothing in Biology," 3.

54. Levene, Ehrman, and Richmond, "Theodosius Dobzhansky," 4; Stebbins, "Variation and Evolution in Plants," 174; Wallace, "Legacies of Theodosius Dobzhansky," 44.

55. Ayala, "Nothing in Biology," 3; Glass, introduction to *The Roving Naturalist,* 2.

participated in the synthesis met with historians and philosophers of science. After the meeting, Mayr proclaimed: "There is complete agreement among the participants of the evolutionary synthesis as well as among historians that it was one particular publication that heralded the beginning of the synthesis, and in fact was more responsible for it than any other, Dobzhansky's *Genetics and the Origin of Species* (1937)."[56] Gould recalls a similar agreement among the founders of the synthesis at the end of the meeting:

> When they could be drawn back into their pasts, one distinct memory emerged. The great works of the synthesis had not been a set of independent volumes, each drawing its separate inspiration from Fisher, Haldane, and Wright. Instead, all the gathered authors looked at Dobzhansky (who clearly enjoyed the accolades) and said that they had drawn primary inspiration from *Genetics and the Origin of Species* (1937). Dobzhansky had not simply been first, by good fortune, in an inevitable line; his book had been the direct instigator of all volumes that followed.[57]

Geneticists and naturalists alike were influenced by the book. According to the cytogeneticist Hampton L. Carson, who recalls reading the book when he was in graduate school, "the excitement that it immediately engendered throughout biology defies simple description and is worthy of historical study in its own right."[58] The geneticist Leslie C. Dunn says that it was Dobzhansky's book (and the lecture series from which the book was drawn) that influenced him to apply genetical ideas to evolutionary problems.[59] Its impact on naturalists can best be seen by examining the other works that represent the influential literature of the synthesis period, for example, Ernst Mayr's 1942 *Systematics and the Origin of Species,* G. G. Simpson's 1944 *Tempo and Mode in Evolution,* and Ledyard Stebbins's 1950 *Variation and Evolution in Plants.*[60] Each of these authors is considered an architect of the synthesis, and each looked back to Dobzhansky's book as the inspiration for his own work. Mayr says that he was "delighted" with Dobzhansky's book, wherein he learned about the mathematical population genetics that was so crucial to his own understanding of the synthesis.[61] Simpson makes an even stronger acknowledgment of the influence Dobzhansky had on his own writing:

> My own thinking along theoretical lines was nevertheless mostly along lines of historiography and organismal adaptation, in fossil

56. Mayr, *Growth of Biological Thought,* 569.
57. Gould, introduction to *Genetics and the Origin of Species,* xxi.
58. Carson, "Cytogenetics," 90.
59. Quoted in Provine, "Origin of Dobzhansky's *Genetics and the Origin of Species,*" 101.
60. Glass, introduction to *The Roving Naturalist,* 2; Ayala, "Nothing in Biology," 4; Smocovitis, "Unifying Biology," 29.
61. Mayr, "How I Became a Darwinian," 420–21.

> and recent organisms, until the first edition of Dobzhansky's *Genetics and the Origin of Species* (1937). That book profoundly changed my whole outlook and started me thinking more definitely along the lines of an explanatory (causal) synthesis and less exclusively along lines more nearly traditional in paleontology.[62]

Stebbins suggests that this reaction was a common one among readers of the text:

> [Dobzhansky's book] represents the birth of the modern synthetic theory of evolution. In addition, it attracted the attention of many biologists trained in disciplines quite different from his own, who then extended the synthetic theory in a variety of different directions. The present author was stimulated to apply the theory to plants.[63]

If we take these claims of influence seriously, a close examination of Dobzhansky's founding treatise could do much to explain the factors responsible for the emergence of the evolutionary synthesis. Recognizing this, some scholars have commented on the manner in which Dobzhansky's book helped explain the new theory, disseminate the new information, alter scientists' conceptual patterns, and offer the social motive required to construct this interdisciplinary treaty.

With regard to theory, Dobzhansky did not introduce his own new laws of evolution. According to Gould, the main point of Dobzhansky's book was not to make an argument about any particular theoretical position: "Dobzhansky's original probe toward synthesis was more a methodological claim for knowability than a strong substantive advocacy of any particular genetic argument."[64] Though the treatise did make some claims about evolutionary theory, the most theoretically significant parts of the book were taken almost directly from Sewall Wright's 1932 paper on mathematical population genetics.[65]

Rather than develop new theories about evolution, Dobzhansky popularized theories that were tucked away in isolated disciplinary communities. For the first time, the theories of the Russian school of ecological genetics and of the mathematical population geneticists were brought together and introduced to the rest of the biological community.[66] As outlined above, these theories were instrumental to the construction of the synthesis, but the synthesis did not form when these theories were first developed. For most biologists, the complex mathematical models of the theoretical population geneticists were incomprehensible.[67] It took

62. Quoted in Mayr, "G. G. Simpson," 456.
63. Stebbins, "Variation and Evolution in Plants," 174.
64. Gould, introduction to *Genetics and the Origin of Species*, xxvii.
65. Eldredge, *Unfinished Synthesis*, 14, 35; Provine, *Sewall Wright*, 342, 346.
66. Glass, introduction to *The Roving Naturalist*, 2.
67. Lewontin, "Theoretical Population Genetics," 58.

Dobzhansky to translate the mathematical models of Chetverikov, Fisher, Haldane, and Wright into a book that biologists from other disciplines could read and understand.[68] Dobzhansky himself believes that this was the role his treatise played in the synthesis:

> The reason why the book had whatever success it did was that, strange as it may seem, it was the first general book presenting what is nowadays called . . . "the synthetic theory of evolution." I prefer to call it "biological theory of evolution." I certainly don't mean to make a preposterous claim that I invented the synthetic or biological theory of evolution. It was, so to speak, in the air. People who contributed most to it I believe were R. A. Fisher, Sewall Wright, and J. B. S. Haldane; their predecessor was Chetverikov. What that book of mine, however, did was, in a sense, to popularize this theory. Wright is very hard to read. He has a lot of extremely abstruse, in fact almost esoteric mathematics.[69]

In contrast, reading Dobzhansky was vastly easier.[70] *Genetics and the Origin of Species* somehow refashioned the difficult formulations of the population geneticists, putting those theories in language that could be understood by other biologists.[71]

Dobzhansky's text also popularized the work of others by making new data about evolution available to those who otherwise would not have encountered them. According to Dunn, a geneticist, the book was all about the evidence that was derived when scientists "take the mathematical theory and test it by observations on populations in nature." This evidence worked to convince geneticists of the truth of "the main factors in evolution: mutation, natural selection, sampling in small populations (which Wright had called random genetic drift) and forces which tend to change populations with reference to each other, such as migrations."[72]

Mayr, a naturalist, agrees that the main influence of the book came from its popularization of data, but he sees that influence working on the naturalists. He argues that one of the major roadblocks to cooperation between naturalists and geneticists was the fact that the naturalists had misconceptions about the experimental claims of geneticists. Geneticists in the heyday of the Mendelian period, from 1900 to 1908, wrote books that publicized their findings, and most of these described experimental results that widened the split between naturalists and geneticists. The next generation of geneticists produced results that fit much better with the naturalists' findings; however, none of the later geneticists wrote books or

68. Provine, "Role of Mathematical Population Geneticists," 180.

69. Quoted in Provine, *Sewall Wright*, 345–46.

70. Ibid., 402.

71. Ayala and Fitch, "Genetics and the Origin of Species," 7691.

72. Quoted in Provine, "Origin of Dobzhansky's *Genetics and the Origin of Species*," 101.

made any attempt to counter the naturalists' misconceptions about the state of knowledge in the field of genetics. Consequently, naturalists in the 1920 and 1930s tended to attack genetic programs that no longer existed.[73] What was needed for a synthesis of the two camps was the removal of this barrier, and Dobzhansky's book did much to accomplish this. For example, one of the older geneticist findings was de Vries's discovery that mutation pressure caused evolution in a single generation through "saltational" leaps. By the 1920s and 1930s, geneticists had discovered that most mutations were not the large chromosomal changes that de Vries studied. But according to Mayr, it wasn't until he read Dobzhansky that he learned that most mutations were very small, and therefore more compatible with his own understanding of evolutionary theory.[74] When it came to both theory and knowledge, it seems that Dobzhansky's book built the bridge between warring camps by publicizing models and information that were obscure or unavailable to the people who needed them.

Some scholars have also pointed out that Dobzhansky's text may have influenced its readers by providing a new way of looking at things. There are two conceptual insights discussed by historians; one supports Allen's thesis about the need to reduce mechanistic thinking and the other supports Smocovitis's thesis about the need to increase mechanistic thinking. Recall that according to Allen's thesis, one of the barriers separating geneticists from naturalists was the fact that experimental geneticists tended to simplify evolutionary problems by studying populations that were made up of "pure types," or groups of organisms that shared a uniform genotype. This "typological" approach made the genetic problem easier to treat mechanistically. In these "pure lines," however, natural selection could not get a foothold, because there was no variation on which it could operate.[75] Since geneticists did not appreciate the variation existing in natural populations, they underestimated the evolutionary importance of natural selection. *Genetics and the Origin of Species* solved this problem by emphasizing the variation that exists in natural populations.[76] By talking about diversity within populations, Dobzhansky helped geneticists recognize the dynamic complexity of populations and changed their simple mechanistic thinking in this regard.

The book also helped naturalists overcome their conceptual barrier to mechanistic thinking. The introduction of the Hardy-Weinberg formula

73. Mayr, "Prologue," 6–9; Mayr, "Role of Systematics," 126.

74. Mayr, "How I Became a Darwinian," 421.

75. Natural selection needs within-population variety to work. In a population, individuals who are better adapted to the environment survive longer and produce more progeny than individuals who are not as well adapted. When there is no variation, no individual is more or less well adapted to the environment.

76. Mayr, "Role of Systematics," 128; J. Moore, "History of Evolutionary Concepts," 14.

and the description of natural selection as a "mechanism" both helped Dobzhansky accomplish this task.[77] As Smocovitis explains:

> Dobzhansky had drawn heavily, consciously, on the "classical" genetics of the Morgan school, which, in its mechanistic and materialistic nature, most closely resembled "classical" physics. Genetics (and the physical world of the gene) was used as the grounding for Dobzhansky's new "evolutionary genetics." . . . The title of his book published in 1937 reveals this grounding: *Genetics and the Origin of Species* offered a framework that brought together the material basis for evolution, determined first through the work of geneticists, with causo-mechanical explanations—made mechanical through the models—for evolution.[78]

Finally, with regard to the social level of analysis, several scholars have argued that the reason *Genetics and the Origin of Species* was so effective was that Dobzhansky was uniquely positioned in the social structure of science to make the call for synthesis. As a working geneticist, trained in Morgan's lab, Dobzhansky had the social authority to speak to geneticists about their future. Before becoming a geneticist, however, Dobzhansky had been educated in the tradition of natural history in Russia, and throughout his life he conducted the field research of the naturalist. In short, as Jeffrey Powell puts it, his social position allowed him to "appreciate as could few others the different approaches, concerns, and vocabularies of the two groups of biologists."[79]

Because of Dobzhansky's ability to work in the interdisciplinary space between genetics and natural history, his book was uniquely positioned to urge people with different professional motives to alter their career paths. According to Provine, *Genetics and the Origin of Species* "sounded again and again the clarion call" to test the mathematical models in natural populations, and many of its readers were inspired by the book to heed that call in order to advance their careers.[80]

Prelude to a Rhetorical Reading

Taken together, these explanations of what Dobzhansky contributed to the evolutionary synthesis are satisfying, but only to a degree. In describing the theoretical, informational, conceptual, and social requirements for synthesis, and in suggesting that Dobzhansky met those requirements, they provide broad accounts of *why* his book succeeded in building a bridge between the warring camps. But these explanations only make

77. Smocovitis, "Unifying Biology," 20.

78. Ibid., 23–24.

79. Powell, "'In the Air,'" 364. See also Carson, "Dobzhansky and the Deepening of Darwinism," 58; Carson, "Cytogenetics," 90–91; and Gould, introduction to *Genetics and the Origin of Species*, xxiii.

80. Provine, *Sewall Wright*, 347–48; Provine, "Origins of the Genetics of Natural Populations Series," 64;

preliminary and ultimately unsatisfactory claims about *how* his book was able to affect the influence it had. For example, Richard Lewontin suggests that Dobzhansky persuaded "by the sheer force of his personality and reemphasis of things that were in the literature as information but not really in people's heads as knowledge."[81] But Lewontin does not explain how this force of personality manifested itself in the book, or how it worked such a remarkable influence in uniting scientists from competing disciplines. Nor does he explain how Dobzhansky's reemphasis of known materials was designed to attract the interest of multiple audiences. A great many scholars suggest that Dobzhansky's book was easier to read than the texts of the mathematical population geneticists. But these accounts do not fully explain how Dobzhansky was able to simplify the complex mathematics for his audience. To truly understand the way in which Dobzhansky's book achieved its influence, we need a closer examination of the text itself.

Just as laboratory studies of genetic material were able to provide a closer look at the mechanisms underlying the ultimate causes of evolution, rhetorical analysis of Dobzhansky's text can provide a closer look at the mechanisms of influence that lay beneath the ultimate causes of the evolutionary synthesis. A rhetorical analysis of Dobzhansky's book and its intertextual antecedents and responses can identify what it was about the construction of the text that allowed biologists to better understand the mathematical theory—and allowed geneticists and naturalists to recognize information that they had previously been unable to see, alter their conceptual frameworks to create the possibility for collaboration, and agree to share their scarce resources. In the next chapter, a rhetorical analysis of Dobzhansky's *Genetics and the Origin of Species* will take the level of our understanding one stage deeper by examining *how* the text worked on its audience.

81. Quoted in Krimbas, "Evolutionary Worldview," 181.

3

A Text Rhetorically Designed to Unite Competing Fields

If the evolutionary synthesis was a "treaty" between warring camps, then it is likely that the document of negotiation most responsible for that settlement was sophisticated in its use of effective rhetorical maneuvers. In order to work its influence on the synthesis, Dobzhansky's *Genetics and the Origin of Species* had to function on several levels: it had to clearly and convincingly lay out the theory that supported the possibility of collaboration, it had to gather what was already known in different camps and weave those threads into a new and coherent pattern, it had to break down the conceptual barriers to unity between the biological camps, and it had to inspire each side to believe that cooperative action was in its own best interest. In addition, it had to do all this without antagonizing either of the two camps.

This chapter discusses some of the rhetorical maneuvers that can be discovered through a study of Dobzhansky's text in its social and intellectual context. The method of rhetorical analysis employed is a close reading of the book itself and a close reading of the written remarks of naturalists and geneticists who commented on the book in reviews or made reference to it in their own research soon after it was published. The findings are organized in a manner that corresponds to the various types of explanation introduced in the previous chapter. Taking as a springboard the comments of scholars who have created broad-level explanations of *why* Dobzhansky's book had the influence it did, this chapter offers a fine-tuned explanation of *how* Dobzhansky's book used specific rhetorical techniques to simplify theory, relay information, alter conceptual patterns, and encourage social harmony.

Simplifying Theory

Historians of the synthesis have argued that Dobzhansky's book was influential because it communicated the theories of the mathematical population geneticists in a manner that was easy for nonspecialists to understand. Recall that Dobzhansky himself claimed that his simplification of the abstruse, esoteric mathematics of Fisher, Wright, and Haldane made *Genetics and the Origin of Species* important to the synthesis. The significance of this factor is confirmed by the remarks of readers at the time that the book was published. More than half of the book reviews that I could find for *Genetics and the Origin of Species* made specific reference to the "clarity" of the monograph. According to William B. Brierley, a reviewer who described himself as an "ordinary botanist/zoologist" observing genetics from afar, "Prof. Dobzhansky has written a most thoughtful and stimulating book, and he has a capacity of making difficult things clear without falsifying them by oversimplification, and of putting his ideas concisely and interestingly in simple language."[1] This sentiment was shared by the laboratory geneticist Hans Grüneberg.

> The exposition is extremely lucid and will indeed come as a revelation to many readers.
>
> The reviewer was particularly impressed by the clear exposition of population genetics, a subject which has for a long time been an esoteric branch of genetics, bristling with the most terrifying mathematics and fully understandable probably only to a few pundits. Now the door is thrown wide open to all of us, and many difficulties for selectionist interpretation vanish like snow before the sun.[2]

It is possible that Dobzhansky saw the need to write about mathematical population genetics in such a clear and simple manner because of his own limitations as a quantitative scientist. At various times in his career, Dobzhansky acknowledged his failure to understand difficult mathematical arguments. He unabashedly admitted that since he could not understand Wright's mathematical treatments, he would invariably read the introductions and conclusions of the papers while trusting that the mathematical middle accurately connected the two.[3]

Examining Dobzhansky's book, one finds that difficult mathematical formulas are not entirely avoided. There are sections, particularly in chapter 6, that require careful attention to a rather specialized mathematical treatment. It is likely, however, that these sections were heavily influenced by Wright. Before sending his book to the publisher, Dobzhansky sent chapter 6 to Wright for editorial review and received advice on how to make the account less "simplistic" and more representative of recent "advances." He made the suggested changes without protest.[4]

Whether or not the few difficult sections in the book were the sections

1. Brierley, review of *Genetics and the Origin of Species*, 669.
2. Grüneberg, review of *Genetics and the Origin of Species*, 69.
3. Provine, *Sewall Wright*, 346.
4. Ibid., 345.

altered by Wright, it seems that they did not mar the overall impression that Dobzhansky could melt away mathematical problems with his readable style. A close rhetorical analysis of the text shows that the author created this impression by using rather conventional strategies of simplification: he extended his discussion of mathematical material so that he could include definitions and explanations, he included both thought experiments and real-world counterparts when describing the derivation of mathematical models, and he used metaphor and simile to create images of the mathematical models for those who were not as easily stimulated by numbers and formulas.

Since most of his theoretical sections drew heavily on the writings of Sewall Wright, it might be instructive to compare a section of Dobzhansky's book with a parallel section of Wright's work. Wright's famous 1931 article "Evolution in Mendelian Populations" used 74 words to explain the effect of "mutation pressure" on gene frequencies:

> The effects of different simple types of evolutionary pressure on gene frequencies are easily determined. Irreversible mutation of a gene at the rate u per generation changes gene frequency (q) at the rate $\Delta \mathrm{q} = - \mathrm{uq}$. With reverse mutation at rate v the change in gene frequency is $\Delta \mathrm{q} = -\mathrm{uq} + \mathrm{v}(1 - \mathrm{q})$. In the absence of other pressures, an equilibrium is reached between the two mutation rates when $\Delta \mathrm{q} = 0$, giving $\mathrm{q} = \mathrm{v}/(\mathrm{u} + \mathrm{v})$.[5]

Dobzhansky included a similar explanation in his own book, but used 260 words to say the same thing:

> The value of q in Hardy's formula, e.g., the frequency of a gene or a chromosome structure in a population, can be modified by mutation pressure in the wide sense of the term, that is by gene mutations and chromosomal changes. If the change from the gene or the chromosome structure A to the state a takes place at a finite rate, the frequencies q and $1 - q$ must change accordingly. Let the mutation in the direction $A \rightarrow a$ have a rate equal to u; the change in the frequency of A in the population will be $\Delta q = -uq$, where q is the frequency of A. If the mutation in the direction $A \rightarrow a$ is unopposed by any other factor, the population will eventually reach homozygosis for a.
>
> Wherever the mutation is reversible, the change in the direction $A \rightarrow a$ is opposed by the change $a \rightarrow A$. With the rate of reverse mutation equal to v, the frequency of A will change as $\Delta q = -uq + v(1 - q)$. An equilibrium will be reached obviously when the change $\Delta q = 0$. The equilibrium value of q determined by the two mutually opposed mutation rates is therefore $q = v / u + v$. Taking, for example, the rate of the mutation $A \rightarrow a$ to be equal to one in a million gametes per generation ($u = 0.000001$), and the rate of the mutation $a \rightarrow A$ equal $v = 0.0000005$, the equilibrium value

5. Wright, "Evolution in Mendelian Populations," 100.

> for q will be 0.33, which means that 33% of the chromosomes will carry the gene A and 67% the gene a. (*DG*, 124–25)[6]

Dobzhansky's explanation, though not simplistic, is easier for a nonspecialist to follow. In the extra space Dobzhansky devoted to his explanation, he defined what was mutating (the gene A was changed to the mutant gene state a), and he provided an example that plugged numbers into the equation to show his audience what the results would be. This extra description made the passage more comprehensible and more interesting to a reader who was unfamiliar with the subject.[7]

Other tools that Dobzhansky used to amplify his explanations were thought experiments and the description of experimental or field reports that corroborated the results of the mathematical theories. For example, when introducing the evolutionary factor called the "scattering of the variability," Dobzhansky began with an explanation similar to the one described above, where he supplied the equations and described an ideal situation through which the formula could be worked out. He then went one step further by describing a thought experiment in which a bowl containing one hundred marbles was used as a stand-in for the statistical theory being described. Finally, he assured his audience that scientists had conducted the experiment with marbles and found the results that would be expected if the theory is correct (*DG*, 127–30). In the next section, where he talked about the interaction of "mutation pressure" and "the scattering of the variability," he followed a similar pattern. First he applied the mathematical model to various ideal conditions, then he introduced and discussed scientific observations and experiments on wild populations to assure his audience that the models apply to the real world (*DG*, 133–38).

The description of ideal situations and thought experiments is a technique to help an audience visualize the abstract mathematics involved in the theories of population genetics. Another way of increasing interest and comprehension is to create images with metaphoric language. For example, throughout the text Dobzhansky used the language of combat to describe the interaction of evolutionary forces. When first introducing Hardy's formula in chapter 5, Dobzhansky spoke about the equilibrium equation as a battle between evolutionary "agents" who act as "opposing forces": "evolution results when one group of them is temporarily gaining the upper hand over the other group" (*DG*, 124). At the conclusion of

6. Quotations from the three primary texts examined in this book are cited in the text with the abbreviations listed below.

DG: Dobzhansky, *Genetics and the Origin of Species.*

SW: Schrödinger, *What Is Life?*

WC: Wilson, *Consilience.*

7. Carol Berkenkotter, responding to an earlier version of my analysis, pointed out that in linguistic terms, the change represents a shift from a specialist register to a nonspecialist register. Berkenkotter, "Interdisciplinary Interpenetration," 184.

chapter 6, the metaphor was extended as migration "counteracts" isolation and "prevents the approach toward a genetic uniformity," and natural selection "combats" the accumulation of unfavorable genes (*DG*, 186). This metaphor added action to the dry equations and helped the audience understand the dynamics of evolution through language that was familiar. Perhaps more important, it appealed to naturalists with the language of competition so closely associated with contemporary formulations of Darwinian selection, helping them come to believe that mathematical formulas could truly represent the action of species survival.

Another metaphor that played a role in Dobzhansky's attempt to make mathematical population genetics intelligible to his audience was the image of a topographic map of populations and gene combinations. With the help of figures that showed imaginary schematic renderings of mountainous landscapes, Dobzhansky talked about what happens to populations that fall into "adaptive valleys" or climb to "adaptive peaks" (*DG*, 187–91). This topographic metaphor was first used by Wright in his 1932 paper "The Roles of Mutation, Inbreeding, Crossbreeding and Selection in Evolution"[8] (see figure on p. 36). When Dobzhansky heard Wright present that paper at the 1932 Genetics Congress in Ithaca, he found it the most convincing presentation of evolutionary theory that he had ever encountered. It was the one time that he did not find the arguments of Wright "esoteric" and impossible for "common mortals" to understand. In fact, he said he "fell in love" with Wright at that meeting, and thereafter used the ideas presented in that paper for the genetics courses he taught at Cal Tech.[9] Recognizing the power of the metaphorical map, he also used it in his summary of population dynamics at the end of chapter 6 in *Genetics and the Origin of Species.*

The visual representations used in this chapter were taken directly from Wright's Ithaca paper, and the explanations of each "map" were very similar to the explanations that appeared there. The metaphor imagines a "field" of all possible gene combinations, graded with respect to adaptive value. "Peaks" on this field are occupied by populations that are well adapted to their environment; "valleys" are dead zones of extinction. As the metaphoric environment changes (for example, mutation pressure increases, or selection increases, or there is more inbreeding), the adaptive peaks shift and the populations inhabiting them change.

At the beginning of chapter 8, on "isolating mechanisms," Dobzhansky reintroduced this metaphor, reminding his readers that the "interactions of mutation pressure, selection, restriction of population size, and migration create . . . new genotypes, which, in the symbolic language of Wright (1932), occupy only infinitesimal fractions of the potential 'field' of gene combinations." What was added in this chapter was Dobzhansky's claim

8. Wright, "Roles," 356–66.
9. Provine, "Origins of the Genetics of Natural Populations Series," 56–57.

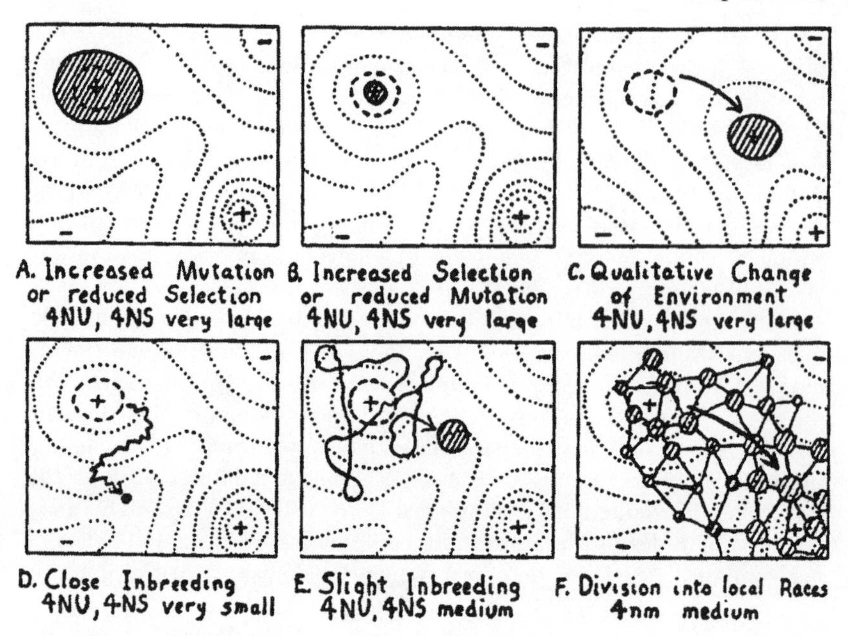

Wright's maps of gene combinations from p. 361 of his Ithaca paper, reproduced in Dobzhansky's *Genetics and the Origin of Species,* p. 189

that *species* are isolated from one another because their "adaptive peaks" are separated by the intervening valleys. "The symbolic picture of a rugged field of gene combinations strewn with peaks and valleys helps to visualize the fact that the genotype of each species represents at least a tolerably harmonious system of genes and chromosome structures" (*DG*, 229). This use of the metaphor extended it beyond its original function in Wright's article, where it was used as an explanation of what happens to a single population that experiences changes in evolutionary pressures. Dobzhansky was now using it to explain why organisms are separated into species that are reproductively isolated from one another.[10]

The rhetorical significance of the map metaphor is found when one recognizes, as the historian Jane Camerini has, that a map can serve "as a conceptual framework, as well as a means of argumentation and, when published, of persuasion."[11] What is probably most interesting about these two uses of the "adaptive landscape" metaphor is the way in which they take a mathematical abstraction of something that is happening in a real environment and return it to the spatial realm with an imaginary

10. Jeffrey R. Powell says that when he asked Wright how he felt about Dobzhansky's extension of his adaptive landscape model, Wright indicated that he did not feel this extension was an abuse of the theory. Powell, "Neo-Darwinian Legacy for Phylogenetics," 72–73.

11. Camerini, "Evolution, Biogeography, and Maps," 727.

scenario.[12] In the first case, biologists who had no trouble visualizing a real population moving to inhabit a new location could visualize the genetic alteration of populations inhabiting the "field of gene combinations." In the second case, the real spatial distance that exists between species that are well adapted to their own ecological niches was made to correspond to the symbolic distance that separates their genotypes.

One effect of this comparison was to make it easier for naturalists to visualize what was happening in the theoretical models of the population geneticists; it even helped some geneticists who were stymied by the complex mathematics of the specialist group of theoreticians within their own ranks. Another effect that isn't as obvious is the influence it had on reducing one of the differences between the conceptual patterns of the naturalists and the geneticists.[13] Recall that the two groups focused on different dimensions of biological matter. Biologists who were used to studying the natural history of wild populations understood that environmental change and geographic isolation altered the phenotype of populations. What they didn't appreciate was the way in which these changes in populations corresponded to changes in gene frequencies. In contrast, geneticists who were used to studying laboratory populations recognized the hereditary importance of changes of gene frequencies; they just didn't appreciate the importance of the environmental and geographical factors that influence natural selection and species formation. Without necessarily *solving* the conceptual problem, the adaptive landscape metaphor began to bring genetics and natural history together by allowing biologists to think about the abstract genetic problem and the spatial realm at the same time. After assimilating this metaphor, the thought patterns of each group would change: whenever geneticists thought about gene frequencies, they would imagine populations moving about in space; whenever naturalists thought about populations inhabiting ecological niches, they would think about the alteration of gene frequencies. In short, Dobzhansky promoted a sort of conceptual chiasmus, in which each side was encouraged to see the world from the perspective of their counterparts.[14]

Surveying the Results of Research

Historians of the synthesis have argued that Dobzhansky's book was influential because it collected the findings of many important studies that

12. David Depew and Bruce Weber make a similar point when they say that with this representational device, "the statistical properties of genetic arrays are transferred to actual populations in real ecologies." Depew and Weber, *Darwinism Evolving*, 294.

13. I realize that this point fits more precisely in the "conceptual change" section of this chapter than in this "simplification of theory" section. I will comment more on this merging of explanations in the final section of this chapter.

14. The term *conceptual chiasmus* is a neologism I have constructed to indicate a rhetorical strategy that promotes the parallel crisscrossing of intellectual space. It was introduced and defined in chapter 1.

were unknown to the majority of geneticists and naturalists. This is confirmed by the responses of readers at the time. The reviewer Charles Kofoid said that the book was particularly timely because of its survey of recent experimental findings in genetics:

> This biological discipline has now reached a stage in its development when a general critical survey of its data, especially of gene mutations and chromosomal changes derived both from experimental breeding in the laboratory and garden and from surveys of related genetic stocks of genera, species, and races in nature, is opportune both for workers in the immediate field and for the wider circle of all biologists interested in the various aspects of organic evolution.[15]

Theodor Just spotlighted the same aspect of the book in his review, concurring with series editor Leslie C. Dunn's comment on the timeliness of this "summary and synthesis of the new experimental evidence":[16]

> Every reader seriously interested in both the progress of genetics and the present status of evolutionism will gladly subscribe to this statement for the book does represent the most thorough-going and actually up-to-date summary of genetical knowledge and its bearing on evolution published so far.[17]

The mathematical population geneticists may have built the theoretical models that were most used during the synthesis, but it was the ecological geneticists who made the experimental findings that supported those theories; and just as the population geneticists required a publicist, so did the ecological geneticists. Before Dobzhansky's book, naturalists thought that the mutations studied by laboratory geneticists were artificial and therefore different from the variations found in nature; likewise, most laboratory geneticists believed that the observations made by naturalists about continuous variation and natural selection were mere speculation, unsupported by experimental tests and therefore unworthy of serious attention. Before the two groups could fruitfully combine their efforts, they would have to dispel their doubts about each other. The findings of the ecological geneticists, gathered and explained in Dobzhansky's text, served this function by demonstrating the results that are possible when the practices of the geneticists and naturalists are combined. The data collected and analyzed by researchers who were engaged in the genetic study of natural populations was solid proof that the two disciplines were not incompatible.

The rhetorical strategy that Dobzhansky used to introduce these data is called prolepsis, or "anticipatory refutation," in which an opposition

15. Kofoid, review of *Genetics and the Origin of Species*, 549.
16. Dunn, foreword to *Genetics and the Origin of Species*, xiii.
17. Just, review of *Genetics and the Origin of Species*, 1105.

argument is anticipated and directly countered by the rhetor. It works especially well when the opposition argument is a belief that is actually being entertained by the audience; arguments against the rhetor's position arise in the minds of audience members, so the rhetor anticipates and voices those arguments to show that he or she understands the audience's concerns, then he or she refutes those arguments. Dobzhansky used prolepsis to bring out into the open, and then refute with data from experiments, the arguments against a new evolutionary synthesis that would be offered by skeptical naturalists and geneticists.

Addressing Skeptical Naturalists

Dobzhansky directly addressed many of the naturalist's concerns about the artificiality of laboratory mutations in chapters 2–4. He did this first by discussing, in chapter 2, some common "misconceptions" about the evolutionary importance of mutations. For example: "Another type of criticism advanced against the mutation theory asserts that the mutations observed in Drosophila and in other organisms produce deteriorations of viability, pathological changes and monstrosities, and therefore can not serve as evolutionary building blocks" (*DG*, 20). For those who believed that mutations seen in the lab were too discontinuous, representing mutational "monstrosities" rather than the continuous variation seen in the field, Dobzhansky responded with multiple examples of studies (most by R. W. Timoféeff-Ressovsky or by Dobzhansky himself) that proved otherwise. These studies reminded the naturalist that the quality of the mutation can only be determined by the environment in which it finds itself, and that many "harmful" mutations are beneficial to the organism under the right conditions. Dobzhansky continued this line of argument by offering an explanation for why mutations that improve viability are rare. He then provided a list of mutations seen in nature that are adaptive for insects but would surely be considered "monstrosities" if they were found in the laboratory. Finally, Dobzhansky concluded his prolepsis by countering those who claimed that mutations only produce "monstrosities" that would never survive in the wild; he outlined experiments by Timoféeff-Ressovsky and E. Baur that indicated that both small and large mutations are found in wild populations.

In chapter 3, titled "Mutation as a Basis for Racial and Specific Differences," Dobzhansky voiced the belief of "a number of writers" that "speciation is a normal and continuous process . . . apparently always adaptive," whereas "mutation is an abnormal and irregular mode of origin, which . . . is not essentially an adaptive process" (*DG*, 40). He then devoted attention to convincing any naturalist who might hold this belief that "individuals resembling mutations obtained in the laboratory have been repeatedly found in natural populations" (*DG*, 41). Here he discussed experiments on natural populations, many conducted by the Russian school of ecological genetics. These experiments showed that mutations are present in wild populations; that those mutations are small and continuous; that they are geographically distributed; and that they are responsible for individual, racial, and species differences. In chapter 4, Dobzhansky made the same argument about chromosomal changes, showing that the translocations, inversions, deficiencies, and duplications found in laboratory experiments are all found in wild populations (*DG*, 73–117).

The naturalist-historian Charles Kofoid was apparently influenced by Dobzhansky's use of prolepsis; he found this part of the book important enough to comment in his book review that Dobzhansky's "analytical survey of the evolutionary implications of the data of genetics tends to verify the basic significance of the gene, of gene mutations and chromosome changes."[18] Years later, Ernst Mayr claimed that he learned from Dobzhansky's book that many, if not most, mutations are very small, and this worked for him as one of the "major bridges between the thinking of the naturalists and that of the laboratory geneticists."[19]

Addressing Skeptical Geneticists

For the geneticist, Dobzhansky had to dispel the misconception that naturalists' observations were mere speculation. For example, to convince the geneticist that the naturalist's focus on continuous variation was not misguided, Dobzhansky had to prove that this sort of variation was not different in kind from the discontinuous genetic mutations studied in laboratories. To do this, he described experiments with natural populations that showed a Mendelian basis for such variation. In chapter 2, Dobzhansky reminded his readers that phenotypic characteristics are the result of both genotype and environmental factors (*DG*, 15–16). Then in the next chapter, he revealed that "laborious progeny tests" have been conducted on continuously varying wild populations and that these tests "reveal the Mendelian segregation" that had been "obscured" in the phenotype by environmental effects (*DG*, 55–60).

In chapter 6, Dobzhansky sought to remove the geneticist's doubts about natural selection, arguing that it must be more than just a hypothesis in the imagination of the naturalist since it is demonstrated through experimental study. For example, in the section of this chapter titled "Experimental study of adaptation," he explained experimental reports by W. Sukatschew, Timoféeff-Ressovsky, R. Goldschmidt, and Dobzhansky himself that suggested selective adaptation by natural populations (*DG*, 151–55).

The Politics of Synthesis

It is interesting to note that Dobzhansky's strategy for removing misunderstanding may have additionally worked on a political level to assure each

18. Kofoid, review of *Genetics and the Origin of Species*, 551.
19. Mayr, "How I Became a Darwinian," 421.

side that he supported their core beliefs. Geneticists were assured that the mutations they studied in the lab had been shown to be the very wellspring of evolution in wild populations, and naturalists were assured that experiments supported their theory of natural selection. One naturalist read Dobzhansky's survey of experimental evidence with a tone of triumph. After beginning his book review with a comment on the "aggression" that the "dominant experimentalists" were directing toward the small "relict populations of systematists," Karl P. Schmidt remarked that "it is interesting to find that the variation of animals and plants in nature is found by geneticists, after an exhaustive survey of the problem, to be very exactly of the kind postulated by Darwin."[20]

As a technique to remove doubts and begin to build bridges between long time opponents, Dobzhansky's use of prolepsis to counter misconceptions on both sides was well designed. Supporting his arguments with experimental data on wild populations, Dobzhansky persuaded the naturalist that the mutations studied by geneticists were not laboratory induced irregularities, and he persuaded the geneticist that the continuous variation required for Darwinian natural selection was based on mutational change. In addition, by showing that productive work between disciplines was already underway, he indicated that it was possible for his readers to undertake such work themselves.

Using Language That Promotes Conceptual Change

One of the most important requirements for any sort of evolutionary "synthesis" was to convince Mendelians and Darwinians that their respective scientific paradigms were not incompatible. So it comes as no surprise that Dobzhansky set forth arguments for the merging of Mendelian mutation and Darwinian natural selection. In 1938, the naturalist Karl Schmidt claimed that Dobzhansky's reconciliation of mutation and natural selection was one of the high points of the book, and the geneticist Raymond Pearl began his review of the book with a discussion of Dobzhansky's union of the two processes.[21] The rhetorically significant question about this aspect of Dobzhansky's influence is *how* the reconciliation was actually accomplished in the text.

Statics and Dynamics

One way readers of Dobzhansky's book made the conceptual leap that linked Darwinian natural selection and Mendelian genetics was by accepting a semantic distinction that he made between "evolutionary statics" and "evolutionary dynamics." He introduced this distinction early in the book, in a section that was summarized in detail by two of the

20. Schmidt, review of *Genetics and the Origin of Species*, 51.
21. Ibid.; Pearl, review of *Genetics and the Origin of Species*, 211.

reviewers.[22] Here Dobzhansky argued that evolution can be divided into two parts: "statics," which are the factors bringing about changes in the genetic composition of populations, and "dynamics," which are the interactions of these forces in the formation and disintegration of species and races (*DG*, 12). Continuing his description of this new classification scheme, Dobzhansky suggested that "evolutionary statics" occupies the first level of the evolutionary process, where "mechanisms of evolution" such as mutations and chromosomal changes "constantly and unremittingly supply the raw materials for evolution." "Evolutionary dynamics" occupies the second level, where the "influences of selection, migration, and geographical isolation then mold the genetic structure of populations into new shapes" (*DG*, 12–13). Later in the book he confirmed that evolutionary statics, or the origin of variation through gene changes, is the "purely physiological, and in the last analysis physico-chemical" level of evolution. Evolutionary dynamics, on the other hand, is on another level; it is the process whereby these gene changes enter into the "field of action" of factors such as selection. Evolutionary statics can be compared to building materials; evolutionary dynamics would be the construction process itself (*DG*, 119–20).

Recognizing that geneticists and naturalists undertake the study of different dimensions of the evolutionary process, Dobzhansky used the semantic distinction between statics and dynamics to show that the dimensions must be combined to create a complete understanding of biology. As he explained later in the book, the geneticist's Mendelian mutations and the naturalist's Darwinian selection are not competing explanations for evolution but two different dimensions of the same process:

> It is hardly necessary to reiterate that the theory of mutation relates to a different level of the evolutionary process than that on which selection is supposed to operate, and therefore the two theories can not be conceived as conflicting alternatives. On the other hand, the discovery of the origin of hereditary variation through mutation may account for the presence in natural populations of the materials without which selection is known to be ineffective. The greatest difficulty in Darwin's general theory of evolution, the existence of which Darwin himself was well aware, is hereby mitigated or removed. (*DG*, 150)

A look at the response of his readers shows that they absorbed the language he used to conceptualize a harmony between Mendelian genetics and Darwinian selection. Reviewers pointed out that gene mutations are the "sources" or "raw materials" for evolution and that these sources enter into the "field of action" of natural selection and other second-level "dynamic" processes.[23] Even scholars today seem to speak through this

22. Just, review of *Genetics and the Origin of Species*, 1105; Snyder, "Modern Darwin," 47.

23. Brierley, review of *Genetics and the Origin of Species*, 668; Pearl, review of *Genetics and the Origin of Species*, 211.

conceptual filter. The philosopher Robert Brandon argues that the bridge between Mendelism and Darwinism was formed when Mendelism was recognized as providing the mechanism for variation on which Darwinian selection could work.[24] The philosopher Richard Burian says that compatibility between disciplines coalesced when the "*mechanisms*" of evolution were taken from population genetics and the "*patterns* and *results*" of evolution were drawn from other disciplines.[25] In large part, it appears that our understanding of an integrated evolutionary theory is built on the distinction between statics and dynamics that Dobzhansky introduced.

Mechanistic Thinking

Another conceptual advance promoted by Dobzhansky's language choices involved the mechanistic thought patterns that Allen and Smocovitis believe were so important to the synthesis. You might recall Allen's thesis that geneticists had to reduce their mechanistic thinking in order to recognize the complexity of real populations and Smocovitis's thesis that naturalists had to increase their mechanistic thinking to bring themselves up to speed with the rest of the scientific world. On first glance, it seems that Dobzhansky's "statics-dynamics" split would work against both these requirements. The distinction originally comes from classical physics: the word *statics* implies a halted system, simple in its frozen state; in contrast, the word *dynamics* implies motion in all its complexity. Since *statics* was applied to the subject matter of geneticists (for example, the "mechanism" of mutation), it would seem to suggest a continued mechanistic simplification of that discipline. Likewise, the word *dynamics* applied to the naturalists' subject matter (for example, natural selection) would seem to suggest a continued rejection of simple explanation in favor of a focus on complex "processes." Insofar as Dobzhansky was arguing that the two fields must *both* exist to fully explain evolution, he was acknowledging that each could continue in its separate mechanistic and antimechanistic practices. But though his initial description of these two fields used terminology that fit the stereotype of a mechanistic genetics and a dynamically complex natural history, he did not perpetuate those stereotypes. In fact, language choices in other parts of the text seemed to subtly undermine the stereotypes and urge each discipline to change its style of thought.

In support of Smocovitis's thesis that naturalists would have to be impressed with the mechanization of their subject matter for the synthesis to occur, a close reading of Dobzhansky's text shows that the word *mechanism* was used to describe natural selection throughout the text. Natural selection is the probable "mechanism" for adaptation, it is the "mechanism" preserving the balance between living matter and the harsh environment, it is the only known "mechanism" that could cause the observed

24. Brandon, "Introduction," 110.
25. Burian, "Influence," 156, emphasis in original.

morphological effect of mimicry, and it is the sole "mechanism" capable of producing a reconstruction of the genetic makeup of a species population from the existing elements (*DG*, 150, 165, 186). Although natural selection belongs to "evolutionary dynamics," the idea of natural selection as a "mechanism" makes it seem to exist not as a *process*, or "interaction of forces," but as the concrete working *parts* of a machine. Although Dobzhansky never explicitly changed his categorization scheme, when he described natural selection as a "mechanism" of evolution, he seemed to be resituating it as a force in "evolutionary statics," where other "mechanisms" such as mutation are studied through experimental practices. Audience members were thus encouraged to think of natural selection as an evolutionary mechanism that could be taken apart and studied.

This ultimately would have had different effects on the two audiences. The geneticists would have felt more comfortable about studying natural selection once it was labeled as a "mechanism," and this might have allowed them to more easily cross disciplinary boundaries. But more important, in the terms of Smocovitis's theory, naturalists were being asked to change their thought patterns. Natural historians who normally thought of natural selection as a more complex "process" were being asked to think of it as a mechanism that could be studied in the same rigorous experimental way as other mechanisms.

In support of Allen's thesis that geneticists had to eliminate their mechanistic bias and recognize the complexity of real populations, a close reading of Dobzhansky's text uncovers several passages where the complexity of natural populations was impressed on geneticists. For example, when describing population genetics, Dobzhansky warned that a simplistic understanding must be avoided.

> A population may be said to possess a definite genetic constitution, which is evidently a function of the constitutions of the individuals composing the group, just as the chemical composition of a rock is a function of that of the minerals entering into its make-up. The rules governing the genetic structure of a population are, nevertheless, distinct from those governing the genetics of individuals, just as rules of sociology are distinct from physiological ones, in spite of being merely integrated forms of the latter. (*DG*, 11)

Thus with an analogy that suggests that the whole is different from the sum of the parts, Dobzhansky introduced a nonreductionist reading of biological explanation. If the rules of population genetics are distinct from the rules of genetics proper, then the powerful reductionist techniques of the laboratory scientists would not be sufficient to explain evolution. For the working geneticist, a combination of micro-level genetic analysis (statics) and populational-level analysis (dynamics) would be required to fill out a complex holistic account.

In addition to making the comparison between population genetics and sociology, Dobzhansky used a laboratory simile to alert geneticists to the danger of oversimplifying their understanding of genetic effects. Though one might be tempted to name a gene for a single striking change that it causes, doing so would be "as naive as to suppose that a change of the hydrogen ion concentration is always a 'color gene' because it produces a striking effect on the color of certain indicators" (*DG*, 28). With this simile, the author argued that a mechanistic approach would be crude and misleading when applied to genes and their effects on populations.

In short, through semantic choices that he made in definitions, metaphors, analogies, and similes, Dobzhansky not only encouraged the recognition that conceptual harmony was possible but also worked to make naturalists think more mechanistically and geneticists think less mechanistically. When we recall that Dobzhansky also used an expansion of Wright's topographic map metaphor to make geneticists think more like naturalists and vice versa,[26] we come to recognize just how expertly he negotiated the different conceptual patterns of his two audiences.

Addressing Social Concerns

Before the synthesis, naturalists and geneticists belonged to rival scientific communities, each competing for the same financial resources. As in any conflict, the removal of intellectual barriers was probably not enough to bring about an end to hostilities, let alone usher in an era of collaboration; what was also needed was a positive incentive for cooperative action. In order for Dobzhansky to negotiate an interdisciplinary treaty with his *Genetics and the Origin of Species*, he had to include more than the communication of theoretical, informational, and conceptual advances; he had to convince geneticists and naturalists that cooperative efforts were in their best interest. In short, Dobzhansky had to remain sensitive to the very different social, political, and professional concerns of each group.

Appealing to Geneticists

Inspirational Calls

During the 1930s, geneticists were experiencing great professional success and enjoying a high status in the larger biological community. Their position in academic institutions was secure and growing. They had established their own journals and done a fair job of taking over naturalist journals.[27] Growing since its inception at the turn of the century, genetics was reaping the rewards of a reductionistic, experimental, and mathematical approach to scientific inquiry.[28]

As a relatively new field of study filled with young scientists who were

26. See "Simplifying Theory," above.
27. Smocovitis, "Unifying Biology," 14–15.
28. Sapp, "Struggle for Authority," 334–41.

eager to make their mark in a hot new line of research, genetics was in a position to be shaped by an influential rhetor with a clear vision of the future. Dobzhansky recognized this and used his voice as a respected geneticist to urge his readers toward the study of evolutionary issues that were traditionally studied by naturalists. Peppered throughout his book are challenges to take up particular studies that will potentially solve an "elusive" or "momentous" problem or fill in a "rather neglected" or "virgin" field. These "clarion calls" inspired geneticists by appealing to their desire to succeed as scientists. In addition, Dobzhansky was careful to lead his readers away from the study of genetic issues that have little to say about evolutionary problems.

For example, in the first chapter, three levels of genetic inquiry were described. The first type is "the genetics of the transmission of hereditary characters," which is "by and large, understood now." The second level of genetic inquiry is "developmental genetics," which has a long way to go but is not promising at this point in time because "no reliable methods have been devised for investigations in this field." The third subdivision of genetics is "population genetics," which studies evolutionary mechanisms in natural populations. According to Dobzhansky, it is only this last level that shows promise "in a not too distant future" (*DG*, 10–14).

Expanding on this claim, Dobzhansky made numerous inspirational calls that urged geneticists to study natural populations. For example, he told them that one route to success would be to gather data about the frequencies of individual genes in different geographical regions. "The amount of attention which the problem of races has attracted in genetics is very small, altogether out of proportion to its theoretical and practical importance. . . . Clearly, here is an almost virgin field for future work" (*DG*, 61–63). He made a similar call for the study of gene frequencies in species: "We may conclude that the total number of genes responsible for the difference between a pair of species has in no published instance been accurately determined. This is simply another way of saying that the genetic analysis of the differences between species has been in no case complete" (*DG*, 67).

In addition to inspiring future population geneticists, he called out to geneticists who were interested in other problems of importance to evolution. Speaking to the cytogeneticist, he argued that for the most part, the fundamentals of their field have already been discovered.

> The chromosome theory of heredity in general and the theory of the linear arrangement of the genes in particular have been put to a crucial test, which has proved their coherence. Some problems, mostly bearing on the genetics of the transmission of hereditary characters (see above), were approached and partly solved. Another class of problems, which happens to be more momentous for us in this book, has been so far largely neglected. (*DG*, 80)

Those new problems were the ones bearing on evolution. To get future cytogeneticists interested in one of these problems, he hinted that research on chromosomal repeats has the potential to bear fruit:

> A study of the repeats may throw a light on the elusive problem of the formation of new genes in the phylogeny of diploid organisms, which thus far has defied all attempts to study it. . . . This hypothesis is for the time being only a speculation. But if it is corroborated by further work, the formation of repeats by duplication of some of the existing genes will have to be regarded as an important evolutionary process. (*DG*, 99–100)

That this call had an effect on its audience can be seen in a research article written fourteen years later. In this article, a geneticist cited Dobzhansky's book when offering his hypothesis that a particular mutant was formed through just such a case of duplication.[29]

For the laboratory geneticist wishing to attain power over the engineering of new animal or plant types, Dobzhansky suggested the study of evolutionary isolating mechanisms:

> It is a fair presumption that the pessimistic attitude of some biologists . . . is due to the dearth of information on the genetics of isolating mechanisms. . . . So long as the genetics of the isolating mechanisms remains almost a terra incognita, an adequate understanding, not to say possible control, of the process of species formation is unattainable. (*DG*, 232)

The scientific record provides evidence that this call of Dobzhansky's was heard and acted on by geneticists who read it. In scientific articles written in the twenty years after Dobzhansky's book was published, several geneticists cited his focus on isolating mechanisms in their own attempts to understand the operation of this evolutionary force.[30]

Through inspirational calls like these, Dobzhansky made geneticists believe that it would be in their best interest to study the subject matter and try to answer the questions of naturalists. He implied that power and success were available to the geneticist who entered the territory of natural history.

Genetic Drift

Another rhetorical choice that tapped into the professional interests of geneticists was Dobzhansky's discussion of the nonadaptive force of "random variation in gene frequencies," a factor that was later to be called "random drift" or "genetic drift." In chapter 5, Dobzhansky spent a good

29. Lindquist, "Mutant 'Micro,' " 417.

30. Rendel, "Genetics and Cytology," 288; Lawrence, "Studies on *Streptocarpus*," 16; Valentine, "Studies in British Primulas," 229; Khân, "Pollen Sterility," 129; Suley, "Genetics of *Drosophila subobscura*," 12; Gerstel, "New Lethal Combination," 630.

deal of time explaining how small populations are influenced by the "scattering of variability" in the process of reproduction (*DG*, 127–48). Relying on Wright's analysis of the Hardy formula, he pointed out that random factors can bring gene frequencies to 0 or to 100 percent. "Thus both fixation and loss of a gene in a population may occur without the participation of selection, due merely to the properties inherent in the mechanism of Mendelian inheritance" (*DG*, 131). When populations are divided into local subpopulations, the influence of genetic drift on each of these small breeding communities can result in the formation of races that never experienced the action of natural selection (*DG*, 134). In fact, Dobzhansky claimed that for many cases in which species are divided into small colonies, random variation is a more likely cause of evolutionary variety than natural selection:

> Random variations of the gene frequencies are a much more probable source of the microgeographic races. With the present status of our knowledge, the supposition that the restriction of population size through the formation of numerous semi-isolated colonies is an important evolutionary agent seems to be a fruitful working hypothesis. (*DG*, 148)

Dobzhansky's focus on this type of evolutionary force implied that geneticists did not need to accept Darwinian selection as the *only* driving influence behind the origin of species. Instead, they could accept an alternate theory that relied on a statistical analysis of random forces. This was important to geneticists because they knew they were not well trained to add to the study of natural selection. After all, naturalists develop hypotheses about the selective advantage of particular variations by investigating wild populations and by using their methods of field observation to recognize subtle differences in the ecology of an environment. In contrast, the idea of genetic drift as an evolutionary agent was particularly well suited to the type of methods that geneticists use. Since it did not require the researcher to recognize the selective advantage of a particular trait in a real environment, genetic drift provided an evolutionary focus that the geneticist could call her own.

That Dobzhansky's treatment of genetic drift was influential to geneticists is clear from the book reviews they wrote. All three reviews written by geneticists devoted space to the discussion of this evolutionary force.[31] Although genetic drift only appeared in a part of one chapter in Dobzhansky's book, Laurence Snyder, Raymond Pearl, and Hans Grüneberg devoted respectively, one-fifth, one-third, and one-half of their essays to excited discussions of this hypothesis and what it implied about natural selection. On the other hand, of the seven naturalists who wrote book reviews, only one even mentioned the fact that nonadaptive evolution was

31. Snyder, "Modern Darwin," 47; Grüneberg, review of *Genetics and the Origin of Species*, 69–70; Pearl, review of *Genetics and the Origin of Species*, 211.

discussed in the book.[32] Alfred E. Emerson offered one sentence about the "possibility" of nonadaptive evolution and then belittled its importance with a "however" clause that maintained that the norm is "adaptive divergence" between species, subspecies, and races.

Historians and philosophers of science have puzzled over Dobzhansky's promotion of genetic drift in the first edition of his book and his later abandonment of that nonselective force in subsequent editions.[33] In the early 1930s, genetic drift was a theory popular with Wright but dismissed by the other mathematical population geneticists. According to John Beatty, Dobzhansky included this theory in his first edition because it supported his anti-eugenicist belief that species should have a large amount of variation. Such a high degree variation could not be supplied by the variation reducing force of selection, but could be supplied by geographical races formed through genetic drift. Dobzhansky then dropped genetic drift in later years when his own experiments did not support it and when he found a special type of natural selection that could do the same ideological work.[34]

Though Beatty shows that Dobzhansky had empirical and philosophical grounds for including genetic drift in his first edition and dropping it from later versions, a rhetorical reading of the text and the response of its readers shows that a secondary effect of Dobzhansky's choice to include this theory in his first edition was to inspire geneticists to engage in evolutionary study. It drew geneticists toward the study of the evolutionary forces that are active in producing variation in wild populations. In later editions, after the synthesis was more of a reality, it was not as difficult for Dobzhansky to convince geneticists that they had a place in the study of that level of the evolutionary process. Genetic drift could be abandoned in later editions for empirical and ideological reasons without having a negative influence on the synthesis of disciplines since geneticists at that time were no longer quite as skeptical about their potential to contribute to the study of evolutionary processes.

Of course, naturalists did not simply skip over this chapter when they read Dobzhansky's book; a closer look at Emerson's review shows that the naturalist reader may simply have read it with a different emphasis. One of the implications of accepting genetic drift as a significant evolutionary force was that an investigation of this force would require mathematical and theoretical analysis; but before that could be accomplished, it would require field studies to provide data about the number of breeding individuals in a species and the amount of interbreeding between colonies (*DG*, 138–39). As David Depew and Bruce Weber point out, genetic drift is an important theory to the naturalist because unlike the mathematical

32. Emerson, "Origin of Species," 154.

33. Provine, *Sewall Wright*, 344–45; Gould, introduction to *Genetics and the Origin of Species*, xxxv; Beatty, "Dobzhansky and Drift," 271–311.

34. Beatty, "Dobzhansky and Drift," 271–311.

models of Fisher and Haldane, it moves "ecological parameters," such as population size and migration rate, to the explanatory foreground.[35] The promise that naturalists would be needed to find these parameters was the social interest that attracted naturalists when they read this section of the book. For example, Emerson's review made it clear that he thought natural selection was the norm and genetic drift a minor player. However, he seemed excited about what Dobzhansky's focus on genetic drift implied about the importance of ecological studies of the size and density of natural populations: "Influences of population size and density receives a proper emphasis. Ecologists have recently become aware of the importance of exact study of the biology of populations, but Dobzhansky, largely interpreting Wright's mathematical analyses, shows the importance of these studies from the standpoint of evolution."[36] Naturalists such as Emerson recognized that the study of genetic drift would force geneticists to collaborate with naturalists and would thus preserve a place for naturalists in the future of evolutionary study.

In short, from the response of the two audiences, it appears that Dobzhansky's discussion of genetic drift was polysemous; it was read with different emphasis by geneticists and naturalists, allowing each to see a different social interest supported by the theory. Geneticists saw an evolutionary mechanism that was easier for them to study than natural selection because genetic drift did not require them to recognize the selective advantage of particular traits in real environments. Naturalists looking at genetic drift saw a continuing need for their own exacting skill at examining ecological parameters such as population size and migration rate.

Appealing to Naturalists

The social concerns that would guide the naturalists of the 1930s into synthesis were very different from the ones guiding geneticists. First, it is clear that Dobzhansky was not primarily using the lure of new research territory to persuade the naturalists to collaborate with geneticists. Dobzhansky's call for the study of population size was the only "inspirational call" for new research that he made to the field naturalist. He also offered few promises that collaboration would result in solutions to old research problems. For example, he hinted that genetics might have an answer to an observation long made by systematists and other natural historians (*DG*, 37, 145). In chapters 4 and 7, he spoke about ways in which the field of genetics might provide a tool for discovering phylogenetic relationships, a long-time goal of natural historians (*DG*, 88–114, 195–201, 211–19). And in the final chapter, he offered a "genetic" solution to the

35. Depew and Weber, *Darwinism Evolving*, 284.
36. Emerson, "Origin of Species," 153.

long-standing problem of the definition of species (*DG*, 309–16). But the "genetic solution" to the species problem was universally rejected by the naturalists who reviewed Dobzhansky's book or who referenced it in their own scientific articles, and the other two incentives were virtually ignored in the same literature.[37] This response suggests that the prospect of finding the answer to traditional problems of natural history was not the primary reason that the naturalist came to accept Dobzhansky's proposal for interdisciplinary cooperation. Instead, the most important incentive for naturalists was an appeal to their social concerns about institutional power.

In the minds of naturalists, their status in the larger scientific community had become an issue of great concern. Geneticists and other laboratory workers were not only encroaching on the naturalist's traditional territory, they were pushing out the older scientists and replacing them with researchers who used a more scientifically respected method. Writing from the Field Museum of Natural History in Chicago, Karl P. Schmidt began his review of *Genetics and the Origin of Species* with a glimpse into the hearts of systematists and other naturalists of the time.

> Among the relict populations of systematists preserved by the museum environment from the competition, if not positive aggression, of the dominant experimentalists, there has been some legitimate concern lest their numbers decline to the point at which extinction becomes inevitable. Huxley's famous reference to the systematists as "Hod-carriers of science" (scarcely in character for him) seems to have been an early sign of an indifference toward systematic biology on the part of those engaged in the more experimental or more mathematical sciences, which has prevailed among university biologists for more than a generation. . . . It is especially gratifying, therefore, to find one of the leaders among the modern geneticists expressing an active interest in systematics. In Dobzhansky's chapter headings appear such topics as "organic diversity," "racial and specific differences," "variation in natural populations," "selection," "isolating mechanisms," and "species as natural units," all of which are of first rate importance to systematic biology.[38]

According to Garland Allen, naturalists who were feeling their resources dry up during the 1920s had taken up a "defensive position" against the more favored field of genetics.[39] The historian Joseph Allen Cain argues that by the time naturalists finally recognized their hopeless

37. For book review arguments against Dobzhansky's genetic species definition, see Brierley, review of *Genetics and the Origin of Species*, 668–69; Emerson, "Origin of Species," 153. For articles that directly challenge Dobzhansky's definition, see Meglitsch, "On the Nature of the Species," 49–65; Boyden, "Significance of Asexual Reproduction," 35; Blackwelder and Boyden, "Nature of Systematics," 30; Boyden, "Comparative Evolution," 21; Baker, "Ecospecies," 63; Mayr, "Speciation Phenomena in Birds," 256.

38. Schmidt, review of *Genetics and the Origin of Species*, 51.

39. Allen, "Naturalists and Experimentalists," 195.

position in this "war," they decided that they would much prefer negotiation to unconditional surrender, and consequently, they were quick to support an evolutionary "synthesis."[40] Schmidt seems to affirm this hypothesis when he indicates that at the time Dobzhansky wrote his book, the naturalists were less interested in winning the war than in participating in a modern biological science that retained a place for them. In fact, of all the reviews of Dobzhansky's treatise, those written by naturalists were the only ones that commented on the way the author negotiated the two fields of inquiry. According to the geneticists, this book was simply "a thorough exposition of the implications for evolution arising from modern genetic theory" and a synthesis of the most recent "developments in genetics."[41]

Perhaps geneticists failed to pay close attention to the way in which power relations were treated in this book because they had no need to defend their professional interests. For them, this book was simply an approval of their advance into territory occupied by an outdated community of "hod carriers." On the other hand, the power struggle was an issue of great concern for the naturalists. For example, Ernst Mayr began his short review with words of praise for the peace offering of Dobzhansky the geneticist. "The often-heard criticism of the geneticist that he is only interested in the mechanism of the transmission of hereditary characters from parents to offspring but not in evolutionary problems, is no longer valid."[42] Likewise, William B. Brierley recognized that Dobzhansky's text could be seen as a negotiating tool. After arguing that all biologists must begin to learn about the rapidly developing field of genetics, he highlighted Dobzhansky's capacity to beat a path between the old and the new.

> Yet, genetics is an experimental study of the laboratory and breeding pen, with immediate results and empirical viewpoints, whilst evolution is an observational study of the museum and field, with long scale results and philosophic viewpoints. What is the relation between cytogenetics and natural history; between for example, the immediate micro-evolutionary phenomena of laboratory cultures of *Drosophila* and the macro-evolutionary phenomena of elephant phylogeny in geological time? Can the laboratory methods and phenomena be put to field test or correlated with phenomena in Nature? For any discussion of such problems most of us are dependent upon the more general writings of natural historians or geneticists, and we can only hope that our guides are reliable. Prof. Dobzhansky seems to be equally at home in both fields and a reading of his book gives one the impression that his guidance is trustworthy.[43]

40. See the discussion of Cain's thesis in chapter 2.

41. Snyder, "Modern Darwin," 47; Pearl, review of *Genetics and the Origin of Species*, 211.

42. Mayr, review of *Genetics and the Origin of Species*, 300.

43. Brierley, review of *Genetics and the Origin of Species*, 668.

For naturalists, the most positive incentive that could be offered was the survival of their tradition. Dobzhansky's focus on the subject matter of natural history, and his competence in discussing those topics, convinced naturalists that his book was a proposal for collaboration, not a hostile takeover.

The rhetorically significant question is how Dobzhansky managed to negotiate a treaty between the two groups in a way that allowed naturalists to see a respectable merger of disciplines but that did not dissatisfy the geneticists who felt themselves to be in a position to call for an unconditional surrender of intellectual territory. Why was this book seen as a defender of the naturalist tradition by the naturalists but as a straightforward exposition of genetic advances by the geneticists? The answer is that Dobzhansky's treatment of the battle between disciplines was worded in a way that allowed it to be read differently by the two audiences. By subtly negotiating the social interests of each group, this strategy of polysemy allowed a peaceful solution to an unequal battle over scientific territory.

On one level, Dobzhansky's book seemed to offer an account of the triumph of genetics over natural history. For example, early in the first chapter, he described the differences between the older morphological methods of the naturalist and the more recent physiological methods of the geneticist. Whereas morphological study was described as an "order-creating and historical discipline" that depends on the method of observing, describing, and cataloging regularities, physiological study was described as a "law-creating science" that analyzes mechanisms and forces through quantitative and experimental methods. Systematics and morphology are disciplines that use the older generalizing methods; genetics is a branch of a new exact science. Dobzhansky then made it clear that his book was not concerned with the old methods: "The aim of the present book is to review the genetic information bearing on the problem of organic diversity; it is not concerned with the morphological aspect of the problem" (*DG*, 6–7). With this sentence, it would seem clear that Dobzhansky was proposing to apply the more rigorous scientific methodology of experimental genetics to a subject area that was being mishandled by naturalists too mired in the old methods. However, this sentence makes the positive response of "morphological" reviewers such as Mayr somewhat mysterious. Why would some naturalists read this book as a cooperative intertwining of two fields rather than as a hostile takeover of evolutionary questions by genetics?

As the chapter progressed, Dobzhansky continued to set up dichotomies like the one between physiology and morphology. For example, recall the distinction he made between "evolutionary statics" and "evolutionary dynamics." The former concerns the forces that bring about change in the genetic composition of a population, the latter the interaction of forces in the formation and disintegration of races and species. Dobzhansky implied that the former is a "first level" evolutionary process "governed entirely by the laws of the physiology of individuals" and the latter is a "second level" evolutionary process governed by "the dynamic regularities of the physiology of populations . . . on which the impact of the environment produces historical changes in the living population" (*DG*, 12–14). Dobzhansky did not favor one side of this division, however, as he had with this distinction between physiology and morphology; instead, he explained that both would be studied: chapters 2–4 would cover the first level "statics," and chapters 5–10 would cover the second level "dynamics" (*DG*, 14).

This was an important point for naturalists in his audience because the "second level process" of evolutionary dynamics was described in a way that seemed to accommodate some of their methods as well as their subject matter. Rather than study laws, the second level of evolutionary process looks at "dynamic regularities"; rather than analyze mechanisms and forces, it looks at their interaction; rather than examine individuals in a laboratory setting, it focuses on "living populations" that have experienced "historical changes" produced by the environment. Although evolutionary dynamics is not an exact duplicate of the morphological method, it draws its attitudes more deeply from the naturalist's camp than from the geneticist's. Assured that most of the book was going to be devoted to the discussion of evolutionary dynamics, the naturalist was able to entertain a more sympathetic reading of Dobzhansky's project than might be expected from a reading of the earlier sentence.

The rhetorical strategy in which one meaning is inserted for a dominant audience and another, hidden meaning is inserted for a subordinate audience is called an Aesopian form.[44] Dobzhansky's message was Aesopian—outward support of the more prestigious science was combined with a subtle turn that supported the less prestigious science—and this was a rhetorically successful move. Interdisciplinary study between two partners that were perceived to be unequal was initiated without conflict because text and subtext were capable of supporting different readings by the different audiences. Geneticists saw the author as right-minded in supporting their superior scientific methods and ended up unknowingly accepting hidden contributions from naturalists that they would have otherwise rejected. Naturalists who were used to hearing their study devalued were not surprised to read a typical attack on their methods but found hope in the subtle airing of their professional concerns.

A quick look at the response of naturalists seems to confirm that at least some level of selective reading occurred. Three of the reviews written by naturalists included a direct quotation of Dobzhansky's stated purpose,[45] but rather than use the sentence cited above, they chose a passage

44. See K. Jamieson, "Cunning Rhetor," 73–78.

45. Brierley, review of *Genetics and the Origin of Species*, 668; Just, review of *Genetics and the Origin of Species*, 1105; Hrdlicka, "Genetics," 240.

that did not so harshly reject their methods: "The present book is devoted to a discussion of the mechanisms of species formation in terms of the known facts and theories of genetics" (*DG*, xv). This rendering of the book's thesis statement balanced the contributions of naturalists (who study dynamic "mechanisms of species formation" such as selection and isolation) and the geneticists (whose "known facts and theories," not methods, were being promoted). Although these naturalist book reviewers might have just as accurately used the other passage, and argued that the book was another misguided attack by a geneticist who failed to appreciate the contributions of natural history, they awarded Dobzhansky high praise for his enlightened and balanced treatment.

Two other intertextual sources should also be noted. First, there was one naturalist who did not read Dobzhansky's book quite as charitably as the others. The ecologist Alfred E. Emerson complained about Dobzhansky's failure to devote enough space to the ecological aspects of evolutionary problems (he estimated that only one-third of the book covered ecology and argued that the discipline's value warranted more).[46] He also read Dobzhansky's treatment of dynamic mechanisms as biased toward genetics.

> Although the various causative mechanisms of speciation are discussed fairly throughout the book, Dobzhansky tends to include all the inquiry concerning causes under the field of genetics. Just which science should claim the principles of natural selection, isolation, migration and population statistics is probably a minor manner, but it is the hope of the reviewer that ecologists will become more conscious of the importance of their contributions to the subject of evolution and species origin than they have been in recent years.[47]

This reviewer recognized Dobzhansky's conspicuous division of fields; however, he did not seem to appreciate the subtle unification that a more charitable reading allows. Instead, he spoke of the book as if it were offering a challenge along the familiar battle lines. The way in which Emerson read the book suggests that an Aesopian message does not always allay the professional worries of members of the weaker, more defensive group.

The other point of special interest is the fact that in Dobzhansky's second and third editions of *Genetics and the Origin of Species*, he altered the passage where he appeared to reject the naturalist's methods. In each edition, he further increased the explicitness of his unification of natural history and genetics. For example, in the 1941 edition, the passage reads as follows:

> The aim of the present book is to review the genetic information bearing on the problem of organic diversity, *and, as far as possible,*

46. Emerson, "Origin of Species," 152.
47. Ibid., 154.

> *to correlate it with the pertinent data furnished by taxonomy, ecology, physiology, and other related disciplines;* this book is not concerned with the *purely* morphological aspect of the problem.[48]

The italics indicate the text that was added in the second edition, words that made for a better balance between genetics and natural history. In the third edition, the passage was removed entirely and replaced with an even more balanced claim for the unification of disciplinary interests. He still spoke of the difference between "historical" and "causal" approaches to evolutionary problems, but this time he made a very even-handed claim that both have contributed much to our understanding. He concluded the section with an explicit move toward synthesis: "Finally, the most recent developments indicate a trend toward synthesis of what were often divergent historical and causal approaches, and toward emergence of a unified evolutionary biology."[49]

There are two possible explanations for this change in the second and third editions. Dobzhansky may initially have been opposed to the methods of the naturalists, but changed his opinion about the value of naturalist contributions as time passed and the synthesis became a reality. If this is the case, it is probably true that any naturalist who imagined a subtle unification of disciplines in the first edition was reading something not intentionally placed in the text. This hypothesis is highly unlikely, however, since Dobzhansky was originally trained as a naturalist and continued to conduct the field studies of the naturalist.[50] It is hard to imagine what could have caused such a drastic, hypocritical, and short-lived change of his opinion about the value of the naturalist methods. Another possible explanation is that he changed this section as a result of changing social conditions. His first edition had succeeded, the synthesis was a reality, and the "war" between naturalists and geneticists was over. It was no longer necessary for a geneticist to disguise his support for naturalists and their methods when speaking to other geneticists. If this is the case, his subtle unification of fields in the first edition and his increasingly less subtle unification in later editions probably indicate Dobzhansky's rhetorical intentions.

Conclusions

By coming to understand more about *how* Dobzhansky's *Genetics and the Origin of Species* functioned rhetorically, we can be more comfortable with our understanding of its contribution to the evolutionary synthesis. A close reading of this text indicates that the document that did so much to negotiate the interdisciplinary treaty of an "evolutionary synthesis" included some rhetorical methods that were straightforward and others that were very subtle. The author communicated clearly and effectively to his audi-

48. Dobzhansky, *Genetics and the Origin of Species*, 2d ed., 7.
49. Dobzhansky, *Genetics and the Origin of Species*, 3d ed., 11–12.
50. Glass, introduction to *The Roving Naturalist*, 1.

ence by using definition, thought experiments, and metaphor; he used prolepsis to make arguments that introduced research findings; he provided linguistic constructions and a metaphoric map to alter perceptions; and he negotiated the attitudes and motives of his readers through polysemous strategies such as the Aesopian message. In short, the text included a battery of techniques to convince two audiences that a united biology was both possible and desirable.

As a final note, it should be acknowledged that even though I arranged this rhetorical reading in sections that correspond to the four broader historical explanations of Dobzhansky's influence on the evolutionary synthesis, it has not always been possible to maintain that separation in practice. For example, the "theory" section of this chapter was set up to describe the rhetorical construction of clarity in Dobzhansky's explication of the mathematical theories; however, the section ended with a comment about how the "adaptive landscapes" metaphor worked on a conceptual level to unite readers from different fields. The "data" section was set up to describe Dobzhansky's use of information to make arguments; however, it ended with a comment about how his choices operated on a social level to reassure readers from different fields. These entanglements of explanatory segments, while complicating the logical balance of this chapter, are not necessarily a bad thing. Any one explanation is insufficient to reveal the way in which a text like this functions; the chapter's movement between theoretical, informational, conceptual, and social explanations shows that these very different historical causes can peacefully coexist in a comprehensive study of a text's influence.

It is the rhetorical perspective that may do the most to draw these different explanations together. Historians and philosophers of science, in their separate scholarly studies of the evolutionary synthesis, each offer their own theories for the causal influences that shape science. Lindley Darden, a scholar with a realist bent, argues that the synthesis formed because the development of knowledge in many different disciplines had reached a significantly advanced state. Joseph Allen Cain, a scholar who looks at the organizational activity of science, is more excited about the sociopsychological motives that inspired synthesis.[51] The rhetorical perspective of close textual analysis, in its own spirit of interdisciplinary cohesion, works as a sort of glue between such diverse causal explanations, bringing them together at the microscopic level of textual influence. This rhetorical perspective says that Dobzhansky was not the most influential architect of the synthesis simply because he popularized the mathematical theories, or because he reproduced the experimental data, or because he helped people conceptualize their fields differently, or because he smoothed political tensions between the disciplines; he was influential because *his text was effectively constructed to do all of these things.*

51. See my discussion of these scholars in chapter 2.

Of course, one could argue that a focus on text merely replaces one sort of sweeping claim with another; rather than assert that the evolutionary synthesis was caused by knowledge development or political negotiation or conceptual insight, the rhetorical approach seems to suggest that the synthesis was formed as the result of a single text. This sort of a reductionist conclusion would be imprudent. Although Dobzhansky's text was greatly influential, we must remember that it was not the only force behind the development of an interdisciplinary research cluster in evolutionary biology. We must avoid the impulse to claim too much for the intentional constructions of a single rhetor and a single text, and recognize Dobzhansky's book as a representative sketch of the influences that were at work in the evolutionary synthesis. Metaphor, prolepsis, Aesopian form, and other compositional choices were the building blocks of interdisciplinary influence, but that influence also took the form of theoretical, informational, conceptual, and sociological developments within two scientific camps. The internal components of this text are inexorably intertwined with the external components of its context.

II Erwin Schrödinger's *What Is Life? The Physical Aspect of the Living Cell*

4

The "Uncle Tom's Cabin" of the Molecular Biology Revolution: Assessing the Place of a Text in History

The evolutionary synthesis was one of the most noteworthy examples of interdisciplinary collaboration in the twentieth century, but another interdisciplinary cooperative effort that took place around the same time may have been even more significant to the history of science. By the mid-1950s, a group of physicists and biologists had combined their efforts to create the new field of molecular biology, a powerful area of study that subsequently developed some of the most remarkable advances in modern science.[1] As in the evolutionary synthesis, scientists from a more prestigious, well-financed discipline (in this case, physics) were convinced to apply their methods to the subject matter of a comparatively less successful science (biology), and scientists from the less powerful discipline were convinced to take a look at the methods of the more powerful one.

The historian Donald Fleming describes the "biological revolution" that resulted in molecular biology as a change in the way scientists looked at biological matter.[2] When O. T. Avery and his colleagues demonstrated in 1944 that deoxyribonucleic acid (DNA) is the material carrier of heredity, they hedged their bets and wrote a paper that was circumspect in its conclusions. According to Fleming, the reason for this restraint was that the biological discipline of genetics at that time practiced the manipulation of hereditary traits "without knowing or caring what genes were made of," and this "formalism" made it difficult for scientists such as Avery to draw conclusions about the physical nature of the gene. In contrast, by 1953, in a paper that heralded the birth of "molecular biology," James Watson and Francis Crick drew supremely confident inferences about

1. There are some interesting books on the development of this discipline. For example, see Olby, *Path to the Double Helix;* and Judson, *Eighth Day of Creation.*

2. Fleming, "Émigré Physicists," 153–55.

their own research on the physical nature of the gene.[3] According to Fleming, the reason that Avery found it difficult to make a claim about the physical nature of genes and Watson and Crick could so easily do so was that between 1944 and 1953, scientists had begun to think about biological matter in physical terms.[4]

Watson began his studies as a traditional zoologist and Crick had been trained as a physicist, but by 1953, both were confident that the most productive scientific research of their time would examine the physical nature of living matter. What was it that motivated them (and others responsible for the birth of molecular biology) to break away from more traditional approaches in physics and in biology? What caused them to embark on the detour of professional trajectories that would eventually lead to this merger of fields?

Several historians have offered general explanations for this interdisciplinary activity. Some highlight the psychological reasons that would drive physicists to colonize the discipline of biology during this period.[5] For example, ambitious physicists who believed that quantum mechanics was the final word on the behavior of atoms may have felt that it was time to move on to a field that still had big problems to solve. Another psychological motivation for physicists may have been the deep guilt that many felt about their participation in the Manhattan project; perhaps some sought a new career in the biological sciences because they had a desire to be part of something that was life-affirming rather than destructive. Another motivation discussed by historians is the social and economic incentive that may have influenced biologists to reshape their field in the image of the physical sciences.The social authority of twentieth-century physics was a powerful force that people such as the geneticist H. J. Muller may have wanted to borrow in order to make biology more successful.[6]

But if we move away from general social and psychological explanations and try to offer a more proximate cause for the professional choices of the scientists who built this new discipline, the evidence overwhelmingly points to one text: Erwin Schrödinger's 1944 monograph *What Is Life? The Physical Aspect of the Living Cell.*

3. Judson also mentions the difference in voice between the two papers. See Judson, *Eighth Day of Creation*, 41. Carolyn Miller frames this issue in the terminology of rhetoric when she talks about the *kairos* associated with Watson and Crick's paper, a *kairos* that she claims was missing when Avery was writing. See Miller, "*Kairos*," 310–27.

4. An alternative explanation is offered by Michael Halloran, who suggests that the reason for the difference between the two papers is simply that Avery, a pre-Kuhnian thinker, was not as savvy a rhetorician as Watson and Crick. However, he does not explain how Watson and Crick, who were also writing before Kuhn, were able to achieve the *Weltanshauungen* perspective that was needed to produce a rhetorically successful revolutionary paper. See Halloran, "Birth of Molecular Biology," 76–78.

5. Fleming, "Émigré Physicists," 156–59; Allen, *Life Science*, 197–98; Teich, "Single Path," 276.

6. Keller, "Emergence of Molecular Biology," 396–401.

The Influence of Schrödinger's Text

According to the biographer Walter Moore, "No doubt molecular biology would have developed without *What Is Life?*, but it would have been at a slower pace, and without some of its brightest stars. There is no other instance in the history of science in which a short semipopular book catalyzed the future development of a great field of research."[7] Fleming calls it "one of the most influential scientific books of the twentieth century."[8] The book was widely read when it was first published and continued to be read through the 1950s. It was reviewed everywhere that mattered, was translated into a number of languages, and was highly regarded.[9]

Most scholars believe that Schrödinger's book had its greatest influence on physicists who, on reading the advice of this Nobel Prize–winning physicist, were inspired to emigrate to the field of biology. But some scholars are convinced that its impact on biologists was also strong. Moore says that the book encouraged biologists to think more rigorously, in terms of mathematically formulated and physically testable models, bringing "physics to the attention of biologists as well as biology to the attention of physicists."[10] Evelyn Fox Keller points out that when participants in the "molecular biology revolution" look back to the influences that nudged them along that path, a great many physicists *and* biologists cite Schrödinger's monograph as the most important inspiration for their career choice.[11] And Neville Symonds says that the book formulated its ideas "in such a way that they gave a sense of excitement about the future perspectives of biology to established biologists as well as to nonbiologists, students, and laymen."[12]

Indeed, the testimony of participants is the most impressive evidence of this book's wide-ranging influence. According to Watson, his move from zoology to the physical aspect of genetics was sparked on reading it.[13] "From the moment I read Schrödinger's 'What is Life' I became polarized toward finding out the secret of the gene."[14] Likewise, M. H. F. Wilkins has cited the personal influence of this monograph on his move in the other direction, from physics to biology: "Schrödinger's book had a very positive effect on me and got me, for the first time, interested in biological problems. I think it had the same effect on other physicists. I think one reason for this is that Schrödinger wrote as a physicist. If he had written

7. W. Moore, *Schrödinger: Life and Thought*, 404.

8. Fleming, "Émigré Physicists," 172.

9. Symonds, "Schrödinger's Influence on Biology," 221; Yoxen, "Schrödinger's 'What Is Life?'" 20–21.

10. W. Moore, *Schrödinger: Life and Thought*, 403.

11. Keller, "Emergence of Molecular Biology," 403–4.

12. Symonds, "Schrödinger and *What Is Life?*" 663. See also Symonds, "Schrödinger and Delbrück," 232.

13. J. D. Watson, "Succeeding in Science," 1812.

14. J. D. Watson, "Growing Up," 239.

as an informed macromolecular chemist it probably would not have had the same effect."[15] Crick agrees with Wilkins's assessment of the rhetorical power of the book, arguing that it was written in a way that made it influential both for himself and for others who made the move from physics to biology. "[O]ne has to recognize that it had a very important influence on younger scientists who were considering entering biology. It certainly did on me. . . . The point is that it made the subject seem exciting and gave the impression to novices that this way of thinking about things would be an interesting line to follow."[16] Elsewhere he wrote:

> On those who came into the subject just after the 1939–1945 war, Schrödinger's little book, *What Is Life?*, seems to have been particularly influential . . . the book was extremely well written and conveyed in an exciting way the idea that, in biology, molecular explanations would not only be extremely important but also that they were just around the corner. This had been said before, but Schrödinger's book was very timely and attracted people who might otherwise not have entered biology at all.[17]

Seymour Benzer was another physicist who turned to the study of genetics after reading Schrödinger's monograph: "I'd always had a latent interest in biology, but it was particularly Schrödinger's book that turned me on. About 1946."[18] The physical chemist Erwin Chargaff says he was "deeply impressed" by the little book. Schrödinger's talk about a "hereditary codescript" excited him because he believed that "the builder's craft" of genetic manipulation could be engaged once that code was broken. Reading Avery's paper at about the same time that he read Schrödinger's monograph, he was inspired to turn his attention to DNA.[19] The chemist-turned-biologist Maynard Olsen says that Schrödinger's book encouraged him to make his career move.[20] And the physicist-turned-biologist Gunther Stent says that he was "captivated" by Schrödinger's book. According to Stent, the book's "propagandist impact on physical scientists was very great":

> [H]aving one of the Founding Fathers of the new physics put the question "What is Life?" provided for them an authoritative confrontation with a fundamental problem worthy of their mettle. Since many of these physical scientists were suffering from a general professional malaise in the immediate post-war period, they were eager to direct their efforts toward a new frontier which, ac-

15. Quoted in Olby, *Path to the Double Helix,* 247. See also Wilkins, "Molecular Configuration," 127.

16. Quoted in Olby, "Schrödinger's Problem," 146.

17. Crick, "Recent Research," 184.

18. Quoted in Judson, *Eighth Day of Creation,* 272. See also Carlson, *The Gene,* 165, 201; and Weiner, *Time, Love, Memory,* 42–45.

19. Chargaff, *Heraclitean Fire,* 85–86.

20. Maynard Olsen mentioned this in a conversation with me in Seattle, Washington, on 26 January 1999.

> cording to Schrödinger, was now ready for some exciting developments. In thus stirring up the passions of this audience, Schrödinger's book became a kind of "Uncle Tom's Cabin" of the revolution in biology that, when the dust had cleared, left molecular biology as its legacy.[21]

Despite the fact that all these scientists speak in unison about the influence of Schrödinger's book, there are a couple of scholars who have taken a more skeptical view about its effect. For example, Neville Symonds, another physicist-turned-biologist who contributed to the molecular revolution in biology, rightly points out that "it is virtually impossible to believe most of what one is told about things that happened 30 to 40 years ago."[22] For him, the claim that the book was influential in the early years of molecular biology is a myth; however, he does think the book was extremely influential on scientists who entered the field *after* the famous Watson and Crick paper.[23] The biochemist Seymour Cohen is also unsure about the book's immediate influence; he says that he does not remember Schrödinger's book being discussed at the summer course on phage biology, a yearly gathering that was to have a great impact on the birth of molecular biology.[24]

When we turn away from corruptible memories and examine contemporary evidence, however, we find that participants in the molecular biology revolution who claim a strong and early influence from Schrödinger's book are not radically reconstructing their memory. In 1979, when the physicist J. R. Reitz remembered that he and his colleagues were excited by the book in graduate school (1946–49),[25] he was not participating in an *ex post facto* bandwagon effect, because in 1950 he co-authored an article that tried to answer some of the theoretical questions raised by

21. Stent "Waiting for the Paradox," 3. See also Stent, "Molecular Biology," 392.

22. Symonds, "Schrödinger's Influence on Biology," 224.

23. Ibid., 221, 225–26. It is important to emphasize that Symonds agrees that the development of the discipline in those crucial years *after* the discovery of the double helix required the inspiration of Schrödinger's book. He says that Schrödinger exhibited a "masterly ability to talk about genes and development in terms of molecules, quantum theory, thermodynamics, and the code," and this ability to speak between disciplines "led to biology being viewed in a different perspective by many non-biologists, and by quite a lot of biologists as well." Symonds says that Schrödinger's book was largely responsible for the post–double helix "influx of new blood which led to many of the new initiatives that enabled molecular biology to blossom in the succeeding years." The book "stimulated countless readers and probably helped initiate several seminal lines of research." See Symonds, "Schrödinger and Delbrück," 232–34.

24. Quoted in Yoxen, "Schrödinger's 'What Is Life?'" 18. Cohen also makes the argument that the book was not influential in the development of molecular biology because by the time it was written, many important people were already working on the boundary line between physics and chemistry. This argument does not dispute the effect of the book on the many scientists cited above who were *not* already engaged in boundary work when it came out.

25. Quoted in Yoxen, "Schrödinger's 'What Is Life?'" 20.

Schrödinger's book. In this article, both he and his co-author claimed that "their interest in biology and the application of physics to biology was considerably stimulated by reading Schroedinger's book."[26] Several other scientific articles written in the years immediately following the publication of Schrödinger's book, either by biologists exploring the physical nature of living matter or by physicists venturing into the territory of biology, cited Schrödinger's book prominently and positively.[27] In the fall of 1946, Max Delbrück opened the discussion at a National Academy of Sciences conference titled "Borderline Problems in Physics and Biology" with the assertion that Schrödinger's book had caused them to come together for what would doubtless be the first of many meetings.[28] This documented assertion lends credence to claims that the book was seriously discussed at such meetings. The scores of reviews that were immediately written by biologists and physicists further indicates that leaders in the scientific community took the book seriously. Many of those reviewers urged scientists to take its message of a productive future for physical biology to heart.[29] The fact that the book was purchased by so many suggests that it was widely read, and the generally positive response of its early readers suggests that it worked at least to some degree as a recruiting tool for Delbrück's summer phage course.[30]

In short, the claim that Schrödinger's book had a significant impact on the development of the new discipline is well supported by the historical evidence. In addition, common sense suggests that the scientists who claimed to have been inspired by the book would have had little reason to reconstruct their memories in this particular way. In the 1960s and 1970s, researchers in the relatively new field of molecular biology may have wanted to reconstruct their early influences in order to draw on the authority of a text that would rationalize their professional choices to others. But it will become apparent in the next few pages that Schrödinger's book was widely critiqued during that period (as well as more recently) for its lack of originality and its scientific flaws. If members of a relatively new discipline were trying to invent an artifact that would, in the eyes of historians and their fellow scientists, validate their earlier professional choices, they could do a lot better than cling to Schrödinger's scientifically invalid text. Since it seems to go against their best interest to claim a

26. Reitz and Longmire, "Living Matter," 15.

27. For example, see Manton, "Comments on Chromosome Structure," 471–73; Butler, "Life and the Second Law," 153; Blair and Veinoglou, "Limitations of the Newtonian Time Scale," 71; Stern, "Nucleoproteins and Gene Structure," 945; Auerbach, "Induction of Changes," 205; McElroy and Swanson, "Theory of Rate Processes," 348; Spanner, "On 'Active' Mechanisms," 488; Stadler, "The Gene," 811–12.

28. W. Moore, *Schrödinger: Life and Thought*, 403.

29. Yoxen, "Schrödinger's 'What Is Life?'" 19–31. These reviews will be examined more closely in chapter 5.

30. See Stent, "Waiting for the Paradox," 6; Kendrew, "How Molecular Biology Started," 141; Mullins, "Development of a Scientific Specialty," 58.

motivational influence from this book, it is likely that the influence to which they lay claim is genuine.

The fact that Schrödinger's book has been the subject of serious academic critique is interesting because it points to a difference between this inspirational monograph and Dobzhansky's. Though scholars engage in some minor quarrels over why Dobzhansky's book had the effect that it had, they agree that it should be valued as an important contribution to the history of science; in contrast, historians, while largely acknowledging the inspirational effect of Schrödinger's book, hotly dispute its value. Likewise, a dispute over the meaning of Schrödinger's book points to a difference between the way in which historians treat this text and the way they treat Dobzhansky's. Though scholars disagree about why Dobzhansky included a discussion of random drift in his first edition, they agree that the book supported that theory; in contrast, historians who write about Schrödinger's book engage in an intense debate over whether Schrödinger's text really supported a particular theory that was mentioned in its pages. In the remainder of this chapter, I discuss these two academic disputes. In the next chapter, I explain how a close rhetorical reading of Schrödinger's text and of the texts that surround it can help us resolve these controversies, and at the same time, bring us to a better understanding of how an interdisciplinary inspirational monograph works on its audience.

The Value of Untrue, Unoriginal Science

The first dispute arises from a disjunction between the strong positive impact of the book on scientists who were to form the new field of molecular biology, and the absolute failure of the book to measure up to scientific standards.[31] As Gunther Stent puts it: "Just why this book should have made such an impact was never quite clear. After all, in it Schrödinger presented ideas that were even then neither particularly novel nor original." Furthermore, there was a "deceptive" clarity to the book that hid gaps in knowledge and "outdated information."[32]

The standards of originality and truth invoked in Stent's evaluation are often used to question the scientific value of Schrödinger's book. Like Dobzhansky's monograph, Schrödinger's did not present an original theory, nor did it provide new experimental data about the nature of living matter. Just as Dobzhansky's book relied heavily on two articles written by Sewall Wright, most of Schrödinger's book was paraphrased from a pamphlet written in 1935 by N. W. Timoféeff-Ressovsky, K. G. Zimmer, and Max Delbrück (sometimes called the TZD, the "green paper," or the

31. For more on why these standards might be inappropriate when it comes to judging this "inspirational" scientific monograph, see Ceccarelli, "Masterpiece in a New Genre" (master's thesis).

32. Stent, "Waiting for the Paradox," 3, 5.

"Dreimännerwerk").[33] The few important ideas contained in Schrödinger's book that were not derived from this pamphlet were also previously developed by other people. For example, Schrödinger's book compared the genetic material to Morse code, and thus included the important notion of a hereditary "codescript."[34] But according to the historian Robert Olby, at least five other scientists had made similar comparisons before Schrödinger.[35] Likewise, the idea that it would soon be possible to solve the coding problem and the recognition that the structure of the gene was important to this solution were first intimated by H. J. Muller, not Schrödinger.[36] Even the idea of writing an inspirational text to call physicists and biologists together was not unique to Schrödinger, since Muller had made a similar plea in *Scientific Monthly* in 1936.[37]

Although these are serious charges, lack of originality is not a fatal flaw. After all, an unoriginal scientific text could still be considered important as a popularization of ideas that were not as well expressed in other works.[38] What makes Schrödinger's work most suspect in the eyes of some historians and scientists is that he got so many of the facts wrong. A few of his most fully developed arguments turned out to be dead ends, leading scientists who took his advice down the wrong path of research.[39] In addition, he missed some of the ideas that were most needed for the development of the new field. For example, he did not offer the insight that complementarity between molecules could explain the specific attraction of chromosomes, even though Linus Pauling, Max Delbrück, and J. B. S. Haldane had made suggestions of this sort before he wrote.[40] He also ignored the crucial question of how genes are accurately replicated from one generation to the next, even though others had recognized the importance of this problem at the time.[41]

The most serious complaints about his work focus on the mistakes that could have been recognized by anyone familiar with the state of scientific knowledge in 1944. As Stent points out, "at the time of writing his book, Schrödinger's knowledge of the state of genetics was some years in

33. Fleming, "Émigré Physicists," 171; Waddington, "Some European Contributions," 321.

34. Fleming, "Émigré Physicists," 174; Judson, *Eighth Day of Creation*, 244–45; Symonds, "Schrödinger's Influence on Biology," 223; W. Moore, *Schrödinger: Life and Thought*, 398; Sarkar, "What Is Life? Revisited," 632, 633–34.

35. Olby, *Path to the Double Helix*, 246. Sahotra Sarkar argues that there was only one other scientist who suggested the possibility of a chemical code inside the hereditary material; see Sarkar, "What Is Life? Revisited," 632.

36. Carlson, "Unacknowledged Founding," 153, 165.

37. Muller, "Physics," 210–14.

38. Carlson, "Unacknowledged Founding," 153, 169–70; Perutz, "Erwin Schrödinger's *What Is Life?*" 234; Symonds, "Schrödinger and *What Is Life?*" 663.

39. Symonds, "Schrödinger's Influence on Biology," 225.

40. Perutz, "Erwin Schrödinger's *What Is Life?*" 239.

41. Ibid., 243; Olby, "Schrödinger's Problem," 143.

arrears."[42] He did not appear to know that Delbrück had given up the genetic study of flies to engage in more productive work on bacterial "phage" viruses.[43] Even more important, Schrödinger did not seem to know that the Dreimännerwerk had been discredited and that Delbrück himself had already abandoned the model of the gene that *What Is Life?* so boldly claimed was the only possible explanation.[44]

Perhaps a lack of expertise in genetics could be forgiven, especially since Schrödinger, a physicist, apologized in advance for potential errors.[45] But when his use of the laws of physics and chemistry was sloppy, there would seem to be no excuse. According to Olby, "Schrödinger was not abreast of current ideas circulating among structural chemists, biochemists, and X-ray crystallographers."[46] As the physicist Linus Pauling explains, "Schrödinger's discussion of thermodynamics is vague and superficial to an extent that should not be tolerated even in a popular lecture."[47] For example, Schrödinger made much of the entropy question, offering the concept of "negative entropy" to explain why living matter seems to defy the second law of thermodynamics. But his explanation was misleading; the only reason his account shows a difference between the thermodynamics of living and nonliving matter is that he failed to define the system in which the measurements would be taken. Had Schrödinger pointed out that the entropy of the entire universe must be considered when applying the second law, there would have been no thermodynamic problem for him to address. By neglecting the outward flux of disorder from organisms, and introducing the idea of "negative entropy," Schrödinger erroneously made it seem as if living matter was somehow different from nonliving matter.[48] Another bit of sloppy reasoning occurred when Schrödinger raised a thermodynamic question about the stability of the gene itself. In doing so, he completely ignored the fact that enzymes exist, and he failed to recognize that they could be used to solve the problem he described.[49] Finally, when he called the gene an aperiodic crystal he

42. Stent, "Waiting for the Paradox," 5.

43. Ibid.

44. Carlson, "Unacknowledged Founding," 161–63; Keller, "Emergence of Molecular Biology," 402; Teich, "Single Path," 277; Perutz, "Erwin Schrödinger's *What Is Life?*" 237.

45. Schneider, "Schrödinger's Grand Theme Shortchanged," 300.

46. Olby, "Schrödinger's Problem," 137.

47. Pauling, "Schrödinger's Contribution," 229.

48. Ibid., 229–31; W. Moore, *Schrödinger: Life and Thought*, 399; Olby, "Schrödinger's Problem," 127. Some scholars argue that Schrödinger's thermodynamic reasoning was deliberate and perspicacious, introducing the physicist Ludwig Boltzmann's ideas about negative entropy to an audience that had not yet recognized the significance of nonequilibrium thermodynamics. They suggest that scientists who complain about Schrödinger's approach to the thermodynamic question are forgetting the tradition in which he was speaking. See Welch, "Schrödinger's *What Is Life?*" 47; Schneider, "Schrödinger's Grand Theme Shortchanged," 300.

49. Pauling, "Schrödinger's Contribution," 230–31; Perutz, "Erwin Schrödinger's *What Is Life?*" 243.

did not seem to be aware of the fact that organic molecules form covalent bonds, which are different from the ionic bonds formed in crystals.[50]

In short, it is clear that novelty and factual accuracy were sorely lacking in Schrödinger's book. But it is equally clear that the book strongly inspired its audience. So when scholars write historical commentaries about *What Is Life?* these two conflicting facts contribute to three different assessment strategies. Some engage in a reading of the text that emphasizes the flaws; this group argues that though the book motivated many readers, we should not value it because it made no direct contribution to the history of ideas. Others engage in a reading that attempts to recover any intellectually satisfying components; this group argues that we *should* value the book, but only because it really did make a contribution to the history of ideas. Both of these groups look closely at the content of the text, but they ignore its inspirational effect when judging its value to the history of science.[51] The final group believes that the inspirational effect of a text *should* be considered when evaluating its importance to the history of science; they argue that Schrödinger's book should be valued because it motivated so many scientists to engage in interdisciplinary research. This group ignores the content of the text altogether when they offer their positive assessment, however; they point to the impact it had on the scientists who read it, and when pressed, they attribute that impact to the prestige of the author and the book's fortuitous timing. I argue that none of these three assessment strategies fully accounts for the book's place in the history of molecular biology; what is needed is a study of the text that recognizes its inspirational value but that also conducts a close reading of its content to explain how the text's arguments allowed it to achieve its remarkable effects.

Assessment 1: We Should Not Value the Book

With all the factual errors and lacunae, and with the absence of any truly original ideas that would add to the development of a new discipline, it is hardly a wonder that scientists such as Linus Pauling would argue that a praiseworthy evaluation of the monograph is misguided.

> To what extent, aside from his discovery of the Schrödinger equation, did Schrödinger contribute to modern biology, to our understanding of the nature of life? It is my opinion that he did not make any contribution whatever, or that perhaps, by his discussion of "negative entropy" in relation to life, he made a negative contribution. . . . When I first read this book, over 40 years ago, I was

50. Olby, "Schrödinger's Problem," 137.

51. You could say that these two groups mistake the "history of ideas" for the "history of science"; they value only texts that include theories or data that make direct contributions to the forward march of scientific knowledge.

> disappointed. It was, and still is, my opinion that Schrödinger made no contribution to our understanding of life.[52]

Seymour Cohen agrees with this assessment, concluding that Schrödinger's book does not warrant space in historical treatments of the discipline.[53] Max Perutz shares this attitude when he reluctantly concludes his discussion of *What Is Life?* in a Festschrift to Schrödinger with a negative final assessment: "Sadly, however, a close study of his book and of the related literature has shown me that what was true in his book was not original, and most of what was original was known not to be true even when the book was written. Moreover, the book ignores some crucial discoveries that were published before it went into print."[54]

Scientists who judge the prototypical scientific text by the standards of truth and originality tend to assess Schrödinger's book by the same standards and conclude that a negative evaluation of the book is in order. There are two problems with this judgment, however. First, Schrödinger's book does not belong to the prototypical genre of science; it is an inspirational monograph, and should be judged accordingly. By focusing only on the technical flaws and the unoriginality of Schrödinger's scientific arguments, these critics fail to consider the social influence of the book on the history of molecular biology, an influence that was very important to the development of the field. Second, when they critique the science of Schrödinger's book with the benefit of hindsight, they tend to recognize only the facts that our current understanding of molecular biology designates as important. Although some of Schrödinger's truth claims were known by certain individuals to be false at the time he made them, most were not known to be false by the general scientific community until much later.[55] In addition, some of the claims that we now recognize as wrongheaded may have contributed to the development of the field by offering information that was vital to scientific thinking at the time but that we fail to recognize as important from our current perspective.

Assessment 2: The Book Is Valued as Part of the History of Ideas

The next assessment strategy does nothing to avoid the first problem above, but it does try to avoid the second. For example, the historian Edward Yoxen argues that people who reject Schrödinger's text because it was "wrong" are creating a whiggish history of science that mistakenly judges past arguments on how well they match modern conceptions of scientific knowledge.[56] Yoxen is not ready to praise Schrödinger for his

52. Pauling, "Schrödinger's Contribution," 228–229.
53. Cohen, "Origins of Molecular Biology," 828.
54. Perutz, "Erwin Schrödinger's *What Is Life?*" 243.
55. Sarkar, "What Is Life? Revisited," 633.
56. Yoxen, "Schrödinger's 'What Is Life?'" 28.

unoriginal, untrue science, but he finds redeeming value in the parts of the text that tapped into problems that were considered important at the time. According to him, it is only the distorted lens of our modern understanding of molecular biology that causes us to see Schrödinger's book as a poor contribution to contemporary information theory and the study of macromolecular structure.[57] For Yoxen, Schrödinger's book was not important for its unoriginal popularization of the "informational" code-script idea, nor for its often mistaken accounts of chromosomal structure, but for its treatment of "the problem of biological order," a problem that was widely discussed in the 1940s but is no longer of interest to the scientific community.[58] This problem was not just a short-lived intellectual detour, according to Yoxen, but a conceptual dilemma that had to be resolved before molecular biologists could create the framework they use today in their analytical investigations.[59] Thus, Schrödinger's contribution was important to the conceptual development of molecular biology even though a modern reader of the book might have difficulty getting into the historical frame of mind necessary to recognize this.[60]

Yoxen's rather complex commentary on the value of Schrödinger's text is powerful because it points out a whiggish contamination of the evaluation standards of readers such as Pauling, Cohen, and Perutz. But, like these scientists, Yoxen is so concerned about situating Schrödinger's book in the history of ideas that he fails to even consider the value of its inspirational content. Yoxen says that his reading is powerful because it allows scholars to value Schrödinger's text for its cogent conceptual advances, "despite what some biologists have said about its having been marginal at the level of motivation."[61] Unwilling to settle for the idea that Schrödinger's book worked "merely" as an inspirational force, Yoxen holds tightly to the idea that the book was important for its "deeper" conceptual advances:[62]

> I have been presenting a case for giving serious consideration to *What Is Life?* not only as a specific catalyst in the careers of several eminent scientists, which could be a rather limited role, but also as a more general indication of the ways in which biological discourse has been transformed through the development of molecular biology.[63]

57. Ibid., 31.

58. Ibid., 23–26, 30–31. Four other scholars have also pointed to this aspect of the book: Yourgrau, "Marginal Notes on Schrödinger," 341; Teich, "Single Path," 278; Witkowski, "Schrödinger's 'What Is Life?'" 267; and Welch, "Schrödinger's *What Is Life?*" 47. A somewhat similar argument is made by those who claim that there is current value to revisiting Schrödinger's conceptualization of order in biological systems and negative entropy. See Rosen, "The Schrödinger Question," 168–90, and Elitzur, "Life and Mind," 433–58.

59. Yoxen, "Schrödinger's 'What Is Life?'" 36, 38.

60. Ibid., 42.

61. Ibid., 43.

62. Ibid., 30. See also Yoxen, "History of Molecular Biology," 277.

63. Yoxen, "Schrödinger's 'What Is Life?'" 44.

It seems strange that a historian such as Yoxen would believe that a rhetorical "catalyst" that altered the career trajectories of several soon-to-be eminent scientists should be considered an event of "marginal" or "limited" value in the history of science. This prejudice against the "motivational" content of the book becomes especially problematic when it comes to explaining why the text, which remained in print and unaltered throughout the years, was always so popular with a wide audience.[64] If the main impact of the book on the history of science came from its ability to enter into a technical debate on the problem of biological order, then why was it not abandoned by the scientific community when that problem was no longer meaningful? Why did so many scientists who were *not* involved in the debate on "biological order" buy and read the book from the moment it was published to long after that debate appeared to be over?[65] Though it is probably true that the book had some impact on the debate over biological order, it is also true that the book had an even more powerful motivational effect on those who were to form the new discipline of molecular biology.

Assessment 3: The Book Is Valued for Inspirational Effect

Though scientists such as Pauling argue that this unoriginal, faulty scientific book should not be valued and historians such as Yoxen want us to appreciate the way in which the book was not really "false" or "unoriginal" because it entered into the scientific debates of the time, a third group of scholars suggests that the book should be valued for it social effect. They say that we should not judge the book by the standards of originality and truth, either from a modern perspective or a more accurately historical one. Instead of being seen as an attempt to add directly to the corpus of scientific knowledge, the text should be judged as an attempt to motivate scientists to engage in interdisciplinary research.

When the text is viewed as a contribution on this level, both originality and truth vanish as criteria for judging the text. With regard to the "novelty" criterion, H. J. Muller admits that even though Schrödinger stole ideas from his own inspirational call for the physical study of the living cell, the monograph should still be highly valued since Schrödinger's rhetorical treatment of the problem was more successfully designed to interest the target audience.[66] With regard to the "truth" criterion, the mistakes that were easily recognized by expert scientists such as Pauling seem to have done nothing to alarm those who read the book and were inspired to change their career paths. Schrödinger's faulty reasoning may have led a few scientists down an unproductive alleyway of research, but his call

64. Ibid., 22, 36.

65. Consider, for example, Symonds's claim that the book had its greatest influence in the post–double helix era, when an influx of new blood caused molecular biology to blossom. Symonds, "Schrödinger and Delbrück," 232–34.

66. Muller, "A Physicist Stands Amazed," 90.

for collaboration between physicists and biologists set them all moving in the right general direction. Indeed, though factual precision is a vital component of the prototypical scientific truth claim, the only fact that Schrödinger absolutely needed to get right was the prediction of future success grounding his policy claim that the disciplines should unite to form a new kind of study. As Nicolas Mullins says in his study of Max Delbrück and the "phage group" of proto–molecular biologists,

> leadership and charisma may be the more important factors, much more important, for example, than accuracy in intellectual judgement. . . . [Delbrück] was absolutely right only in his choice of phage as a problem and in his dedication to the goal of discovering the "secret of life." He was quite mistaken as to what that "secret" would look like. Clearly, then, his importance to phage work and the eventual development of molecular biology lies elsewhere than in his intellectual accuracy.[67]

Likewise, Schrödinger was only absolutely right insofar as he chose to dedicate his book to a merger of physics and biology. His importance lies in his ability to motivate scientists to achieve that merger, not in the scientific accuracy of his arguments.

Unfortunately, there is a problem with this third assessment strategy, just as there were problems with the first two. This group of scholars refocuses attention on the motivational impact of the book but does so by retreating from any discussion of the book's content. Most of the scholars who argue that Schrödinger's text should be valued for its inspirational effect offer broad sociological explanations of its influence, using words such as *leadership* and *charisma* to talk about its power. For example, Evelyn Fox Keller argues that the motivational success of the book cannot be attributed in any way to the specific biological arguments that Schrödinger made. Instead, she says, Schrödinger's legacy in molecular biology was made by the timing of the book in the interplay of disciplinary politics and the authority carried by the author's name.[68] This explanation is powerful in that it adds a social component that is missing in the explanations of scholars such as Pauling and Yoxen who restrict their analysis to the history of ideas. But assessments on the broad sociological level offered by Keller do not explain why it was the text that Schrödinger wrote, rather than the texts of other, more credible scientists, that worked the inspirational sensation. If physicists and biologists were simply looking for a *name* to authorize their interdisciplinary alliance, then Delbrück's was available both before and after Schrödinger's book was published. The social authority Delbrück carried was strong since Delbrück not only *said* that physics and biology should merge but believed it strongly enough

67. Mullins, "Development of a Scientific Specialty," 79.
68. Keller, "Emergence of Molecular Biology," 404–6.

to halt his work as a theoretical physicist and enter the foreign territory of the biological sciences. In addition, there was another name that could have just as easily served the same role as Schrödinger's. If boundary-jumping scientists were merely looking for authoritative support, why did they not turn to Niels Bohr, the renowned physicist whose famous lecture on biology was published in the journal *Nature* in 1933?[69]

Although ethos was certainly an important part of the influence of *What Is Life?* there was probably something more to the rhetorical power of this book than simply the fact that a physicist was saying good things about biology. There was something about the text itself that persuaded physicists and biologists to take on a new subject and method of research. A rhetorical reading that asks *why* Schrödinger's text played such a strong leadership role in the development of molecular biology would buttress this third assessment of its value at just the point where it seems to break down. It would follow the social historians in recognizing the book's importance as a motivator of interdisciplinary collaboration, but it would then go on to examine the arguments themselves, to see if and how they contributed to that effect.

Other Laws of Physics

Among scientists and historians who have offered scholarly commentary on *What Is Life?* the second major point of contention concerns a single passage in which Schrödinger claims that his "only motive for writing" is to show that "living matter, while not eluding the 'laws of physics' as established up to date, is likely to involve 'other laws of physics' hitherto unknown, which, however, once they have been revealed, will form just as integral a part of this science as the former" (*SW*, 68–69). This passage has been interpreted in at least three different ways by scholars writing histories of molecular biology, and each group argues that its reading of this passage is the "correct" one because it was what Schrödinger intended (see table on p. 76). The first group argues that Schrödinger meant to invoke Niels Bohr's complementarity thesis, which would imply that new acausal laws of physics would soon be found in the living cell, just as new acausal laws had recently been found in atomic physics. The second group argues that Schrödinger was implying that new deterministic "order-from-order" laws would be found in the biological organism, and those laws would be coherent with classical physics. The third group argues that Schrödinger was not suggesting the existence of "new" laws at all but instead was arguing that our recently developed understanding of those "other," nonclassical, quantum mechanical laws could be directly applied to gene structure to explain living matter. Let us examine each of these interpretations in turn.

69. Bohr, "Light and Life," 311–19.

Stent's Reading Frame	Olby's Reading Frame	Fleming's Reading Frame
New complementary laws of physics will be discovered in biology	New deterministic laws of physics will be discovered in biology	Other quantum mechanical laws of physics need merely be recognized in biology

Three incompatible interpretations offered for Schrödinger's "other laws of physics" passage

New Complementary Laws of Physics Will Be Discovered in Biology

Gunther Stent is the scientist-historian who has most championed the first interpretation, which he believes was "the romantic idea" that fascinated physicists who encountered the book.[70] He believes that in this passage Schrödinger was expressing an idea loosely based on Niels Bohr's 1933 lecture "Light and Life." In this lecture, Bohr asked "whether some fundamental traits are still missing in the analysis of natural phenomena before we can reach an understanding of life on the basis of physical experience."[71] According to Stent, Bohr implied that the answer was yes, and he also implied that those missing traits would not be strictly deterministic in nature. Bohr argued that the physicist studying biological matter would have to admit that there are two complementary sets of physical laws: the laws of physics that operate on inorganic matter and the laws of physics that operate on organic matter. Just as the modern physicist had to accept the fact that light can be described in two "complementary" ways (as particle and as wave) and that atomic motion must be described through both mechanical and quantum physics, the modern physicist would have to accept the fact that the physical laws of nonliving matter were different from, but complementary to, the physical laws of biology. By studying biological organisms and revealing these physical "paradoxes," scientists would be rewarded with a wider and more powerful understanding of natural phenomena that, according to Stent, comfortably navigates "between the Scylla of crude biochemical reductionism, inspired by 19th century physics, and the Charybdis of obscurantist vitalist holism, inspired by 19th century romanticism."[72] In short, the physicist would be able to solve the mystery of life without reducing that mystery to the dull application of already discovered physical laws. Says Stent, "This search for the physical paradox, this quixotic hope that genetics would prove incomprehensible within the framework of conventional physical knowledge, remained an important element of the psychological infrastructure of the creators of molecular biology."[73]

70. Stent, "Waiting for the Paradox," 4.
71. Bohr, "Light and Life," 316.
72. Stent, "Light and Life," 236–37.
73. Stent, "Waiting for the Paradox," 4.

At least six other historical accounts of Schrödinger's book have followed Stent in reading the passage in this way. Max Perutz concludes that "Schrödinger is thus drawn to the same conclusion as Niels Bohr had been, apparently unknown to Schrödinger, 12 years earlier, and one that young physicists found equally inspiring."[74] François Jacob reads the "other laws of physics" passage to mean that Schrödinger, like Bohr, believed that there was a complexity to organic constituents that had nothing in common with classical physics and chemistry.[75] Ernst Mayr believes that Bohr and Schrödinger both turned to a form of vitalism when they "postulated that some day one would discover unknown physical laws in organisms, laws not operating in inert matter."[76] Elof Axel Carlson says that Schrödinger acknowledged his belief in Bohr's view that new physical laws would be needed to interpret life, and that support for this view constituted the "bulk" of his book.[77] G. Rickey Welch argues that Schrödinger was not in favor of reducing biology to physics; instead, Schrödinger was interested in discovering a "romantic" theory that would allow the complementary pieces of those two sciences to fit together.[78] And Garland Allen says that Bohr and Schrödinger were both "antimechanists and antireductionists in the older sense of the words."[79] This meant that they sought "new laws of the physical universe from the study of life," a study that "could possibly open wholly new vistas of the natural world that have been hidden to physicists by the study of strictly inorganic phenomena. As classical physics had to revise its explanatory criteria to account for quantum phenomena, so it might have to revise further criteria to account for biological."[80] According to Allen, this was unquestionably the meaning of the "other laws of physics" passage, which proclaimed that "a new physics was to emerge from the study of biology, just as it had emerged from the study of quantum phenomena."[81]

New Deterministic Laws of Physics Will Be Discovered in Biology

A different reading has been put forward by the historian Robert Olby, who argues that Schrödinger's search for "other laws" must be distinguished from Bohr's search for "new paradoxes." Olby claims that Schrödinger's quest for new laws was an attempt to rediscover in living matter the strict determinism that had been banished from physics ever since quantum mechanics had come on the scene.[82] "He did *not* offer his

74. Perutz, "Erwin Schrödinger's *What Is Life?*" 242.
75. Jacob, *Logic of Life*, 245.
76. Mayr, "How Biology Differs," 46.
77. Carlson, "Unacknowledged Founding," 152, 153.
78. Welch, "Men, Molecules, and (Ir)reducibility," 189.
79. Allen, *Life Science*, 227.
80. Ibid., 224, 196.
81. Ibid., 197.
82. Olby, "Schrödinger's Problem," 136; Olby, *Path to the Double Helix*, 229, 240, 245–46.

readers the bait of a fresh mutually exclusive complementarity relationship, as did Bohr. He offered them 'other laws' to be sure, but not of the kind Bohr envisaged."[83] According to Olby, Schrödinger was asking his audience to find new "order-from-order" laws that would encompass the quantum mechanical "order-from-disorder laws" and thereby recover the causal nature of classical physics. In short, these new laws would be part of a sophisticated theory that could reduce all natural phenomena to relatively simple deterministic physical explanations. By studying the biological organism, the physicist would soon discover the new laws that dissolve the paradox of complementarity and banish the specter of acausal quantum physics.

Several scholars agree with Olby's reading. For example, Neville Symonds says that Bohr looked at biological systems to find the "principle of uncertainty" that would add "a new dimension to ideas about physics." In contrast, "Schrödinger's approach was quite different. He was fascinated by the principle of orderliness which underlined the whole growth and development of living organisms."[84]

Three of Schrödinger's biographers agree that "order-from-order laws" must have been the subtext to the "other laws" passage. According to Wolfgang Yourgrau, even though Schrödinger was one of the founders of quantum mechanics, he was like Planck and Einstein in that he never abandoned the hope that "causality and determinism will some time in the future rule our scientific reasoning again."[85] According to Linda Wessels, this was because Schrödinger was trained under physicists who were taught by Boltzmann. Schrödinger was not "weaned on the paradoxes of Bohr's theory of the atom," and was therefore unable to adapt to the complementarity theory.[86] Ernst Peter Fischer agrees with these assessments and concludes that Schrödinger's search for other laws was the search for a refutation of the acausality of quantum mechanics, a search for "order-from-order" laws that would reintroduce determinism in physics.[87] He writes:

> In his opinion, the inability [of present-day physics to explain biology] is similar to the inability of a man who knows how a steam engine works and who is confronted with an electric generator. Both of these contraptions contain iron and copper, but unless he had studied electrical phenomena the steam engineer will be quite unable to understand the workings of the generator. He might be tempted to believe that the laws of physics break down when applied to the generator, while in reality he is merely confronted with

83. Olby, *Path to the Double Helix*, 245.
84. Symonds, "Schrödinger's Influence on Biology," 225.
85. Yourgrau, "Marginal Notes on Schrödinger," 336.
86. Wessels, "Erwin Schrödinger," 272.
87. Fischer, "We Are All Aspects," 823–24, 828.

> the behavior of matter under a new set of conditions which he has not yet analyzed.[88]

Fischer is here referring to an analogy Schrödinger used in the final pages of his book, an analogy that suggests that the analysis of living matter would soon result in new laws that are consistent with classical mechanics, just as the laws of electrodynamics were consistent with previous mechanistic laws of physics. Although physicists were being tempted to acknowledge that the deterministic laws of physics break down at the quantum level, or that those laws break down in the living organism, this reading of the passage suggests that Schrödinger himself was optimistic that physicists would soon find the new theory that would banish the temptation and reinstate those deterministic laws.

Other Quantum Mechanical Laws of Physics Need Merely Be Recognized in Biology

Olby and his supporters provide a reading of Schrödinger's "other laws" passage that is diametrically opposed to the reading that Stent offers. To make matters even more hermeneutically complex, the third reading is equally incompatible with the other two. This reading was first explained by the historian Donald Fleming. He agrees with Olby that Schrödinger was not fond of Bohr's complementarity thesis.

> Schrödinger never found the idea appealing even as an explication of quantum mechanics, and still less as a kind of philosopher's stone for dissolving all riddles. . . . In fact, he never mentioned complementarity in *What Is Life?* Elsewhere, in responding to a hypothetical inquiry as to why he didn't talk about it, he said flatly that he did not think that it had as much connection as currently supposed with "a philosophical view of the world." In *What Is Life?* he never implied the existence of mutually exclusive biological laws and physical laws, any paradoxical situation where a choice would have to be made between them for the purpose in view.[89]

So like Olby, Fleming counters Stent's reading of the passage. But whereas Olby argues that the reference to "other laws" was a hint that "new order-from-order laws" would soon be found in biological matter, Fleming argues that Schrödinger was really hinting at something else: rather than suggest the need for "new" laws, Schrödinger was saying that his sole purpose in writing this book was to show that the current laws of physics could describe the peculiarities of the living cell. More specifically, when read from this perspective, the passage suggested that the recently discovered quantum mechanical Heitler-London bond forces that hold atoms together in molecules are sufficient to explain the secret of life. "Schrödinger had spoken in *What Is Life?* of 'other' laws of physics, 'hitherto

88. Ibid., 829.
89. Fleming, "Émigré Physicists," 174–75.

unknown,' as required in understanding life. It is at least arguable that the laws he had in mind were those of the Heitler-London forces and that 'hitherto' meant before the application of quantum mechanics to biology by himself and the early Delbrück."[90] Once scientists read his book, they would recognize that the "Heitler-London forces" are sufficient to describe the secret of life, and those laws would no longer be "unknown" by the majority of scientists. An understanding of Schrödinger's own beliefs and a reading of the rest of the book make it most likely "that Schrödinger thought the Heitler-London forces *were* the other laws. For on close scrutiny, the general air of expectancy that every reader recalls as pervading the book, the sense of rising excitement at greater revelations to come, may well have attached for Schrödinger to the logical progression of his own argument. He may have supposed that he was actually answering the question that he asked in his title."[91] According to Fleming, the title question "What Is Life?" was answered in the subtitle, which proclaimed that life is nothing more than the newly discovered (although not yet well publicized) quantum mechanical laws that define the "physical aspect of the living cell."

Like the other two readings, Fleming's interpretation has support from other scholars. The historian John Fuerst points out that Schrödinger "seems to be on firm reductionist ground" throughout the book and argues that Fleming's reading of the "other laws" passage is essentially correct: "Schrödinger was merely a highly sophisticated reductionist."[92] The biographer Walter Moore agrees; he says the ending of the book clearly reveals that any previous hint at "new laws" was actually a reference to the already known Heitler-London forces that make chemical bonds.[93]

Prelude to a Rhetorical Reading

All three readings of the "other laws" passage conflict with one another. In Stent's version, the new laws will be acausal; in Olby's version, the new laws will be deterministic; and in Fleming's version, the laws are not new at all. In the next chapter, I provide evidence that all three interpretations are perfectly well supported by the text. A study that seeks to unravel the history of ideas would try to determine what Schrödinger *really* meant and would be forced to side with one reading rather than another. My rhetorical analysis of the text and its reception offers another way of looking at the problem. Rather than try to discover which reading was the "correct" one, my rhetorical reading recognizes how each of these readings was valid from a certain perspective and served a specific function. Rather that seek the intent of the author, I explain how the polysemy of the passage contributed to the persuasive impact of the book.

90. Ibid., 183.
91. Ibid., 176.
92. Fuerst, "Role of Reductionism," 264.
93. W. Moore, *Schrödinger: Life and Thought*, 400.

In the next chapter, a rhetorical analysis of the text and of the discourse which surrounds it will help us come to understand how this passage and others like it helped inspire biologists and physicists to produce the collaborative field of molecular biology. The value of the book and its meaning will be constructed rhetorically as we come to recognize what it was about the arguments in *What Is Life?* that made the text such a successful case of interdisciplinary inspiration.

5

A Text Rhetorically Designed to Negotiate Different Interests and Beliefs

Those who write the history of molecular biology continue to dispute the value of Schrödinger's book and the meaning of the "other laws" passage. This chapter shows that when the text and its intertextual antecedents and responses are examined from a rhetorical perspective, a solution to both controversies begins to take shape.

The dispute over the value of the book is resolved by focusing on the content that allowed the text to have such a strong motivational influence on its audience. A rhetorical approach acknowledges that Schrödinger's book contains technical errors, making it scientifically flawed from a modern perspective; it also recognizes that the book may have had some historical value in its scientific contribution to the now-forgotten debate over biological order. But instead of focusing exclusively on the book's contribution to the history of ideas, my rhetorical reading understands that the most important role of the book in the history of molecular biology was its influence on physicists and biologists who took up the study of "the physical aspect of the living cell." This reading focuses on the *inspirational* content of the book. Since the scientists motivated by the book were instrumental to the development of molecular biology, the success of Schrödinger's motivational arguments was, in its own way, just as important for science as any new discovery of fact or theory would have been.

This rhetorical focus on inspirational *content* also supplements the readings of social historians who value the book's inspirational influence but who imply that the impact of the book came solely from its timing and the prestige of its author. It is likely that there was more to Schrödinger's text than just good timing and a trusted name; after all, other respected authors made timely attempts at inspiring interdisciplinary integration but failed. Something about the construction of Schrödinger's book allowed it to succeed, something that we can only appreciate when we engage a close reading of both Schrödinger's text and the intertextual documents that surround it.

This perspective also resolves the debate over the "correct" reading of the "other laws" passage. Rather than develop a monologic reading that tries to reveal the truth content of the book, a rhetorical reading that privileges motivational effect focuses on the interaction between the text and its audience. By concentrating on that interaction, we come to understand how the success of the book can be at least partially attributed to the way in which it employed strategies of polysemy to negotiate the contrary interests of its targeted audiences. Since all three readings of the "other laws" passage were instrumental to the persuasive effect of the book, this rhetorical perspective suggests that all three should be recognized and appreciated.

In this chapter, a close rhetorical study cultivates a new understanding of the book and its impact. This analysis argues that much of the success of *What Is Life?* can be attributed to rhetorical techniques that simultaneously appealed to the contrary interests of different audiences. Just like Dobzhansky's successful move toward synthesis, Schrödinger's book contained techniques of persuasion that were instrumental to the development of interdisciplinary collaboration.

Comparison with Other Attempts at Inspiring Interdisciplinary Work

As chapter 4 indicated, most commentators do not address the issue of *how* Schrödinger's book achieved its motivational effect, and when a few scientists and historians do begin to offer explanations, they end up providing reasons that are unsatisfying in their lack of specificity. For example, Crick argues that the book "was extremely well written" and made the subject seem "exciting," and Stent says it worked by "stirring up the passions" of its audience.[1] But neither of them explains precisely *how* it achieved these effects. To understand the influence of the text, we need to understand what it was about the writing that made it so exciting.

Likewise, we need to expand on the broad political explanation offered by Evelyn Fox Keller, who locates the power of Schrödinger's book in the social authority that the world-famous physicist contributed to a collaboration between physics and biology. Keller's explanation, though it certainly captures part of the reason for Schrödinger's success, does not tell the whole story. It does not adequately explain why other powerful scientists who urged collaboration at the time were unable to have the same dramatic effect on audiences. To fully understand the book's inspirational influence, we need to delve deeper than the name on its cover; though

1. Crick, "Recent Research," 184; Crick, quoted in Olby, "Schrödinger's Problem," 146; Stent, "Waiting for the Paradox," 3.

ethos was necessary, it was not sufficient to effect the transformation in disciplinary loyalties that allowed the field of molecular biology to recruit so many workers at that point in history. A close reading of the text and intertext will allow us to more fully understand what made Schrödinger's plea for collaboration more persuasive than other similar pleas.

H. J. Muller

To begin, we must take a closer look at other texts that made the same attempt that Schrödinger made but did not achieve the same success. Keller first develops her argument about the influence of Schrödinger's ethos when she compares his text with a similar call for interdisciplinary collaboration written by H. J. Muller in 1936. According to Keller, the reason that Muller's call went unheeded was that "Muller, being a mere biologist, lacked the authority either to command the attention of physicists themselves, or to persuade other biologists to a program that most of them found totally alien."[2] As a respected geneticist, Muller had the authority to command *some* attention from his *Scientific Monthly* audience, but it is true that he did not carry the same level of authority that Schrödinger did as a Nobel Prize–winning physicist. Nevertheless, a close look at the arguments Muller employed suggests that another reason his call failed was that his text was not well designed to persuade physicists or other biologists to undertake collaborative work.

Muller began his essay on a promising note, suggesting that physicists will find the study of biology important not only because it can elucidate the most fundamental questions of biology but because it might throw light on some of the fundamental questions of physics as well.[3] This would seem to be the sort of appeal that would attract physicists to a new interdisciplinary study. However, it is the only place in the text where Muller considered the professional goals of the physicists in his audience. Rather than write an essay that reduced the barriers to collaboration by speaking to the physicist *and* the biologist about a productive new interdisciplinary study, Muller continually reinforced the boundaries between the two fields by addressing the physicist from the perspective of the biologist. Throughout the essay, Muller used the words *us* and *we* to indicate geneticists who are seeking new answers for biological questions; in contrast, physicists were spoken about in the third person and placed in the position of subordinates who would provide the information that biologists needed to answer those questions. For example, when talking about the gene's property of "auto-attraction," Muller said: "We would like physicists to search the possibilities of their science and tell us what kind of forces these could be, and how produced, and to suggest further lines of approach in their study" (211). When talking about the gene's property of "auto-synthesis," Muller

2. Keller, "Emergence of Molecular Biology," 401.
3. Muller, "Physics," 210.

said: "We would like the physical chemists to work on this problem of auto-synthesis for us" (212). He made similar remarks about how the biologist would have more success in solving biological issues such as growth, reproduction, and heredity "if we had people of sufficient physical training . . . to tackle such a job" (213). Praising the physicist who works on irradiation and mutation as the one who "is to-day most actively and fruitfully helping the geneticist," Muller called on the altruistic spirit of other physicists who were willing to assist their brothers in the biological sciences. "The geneticist himself is helpless to analyze these properties further. Here the physicist as well as the chemist must step in. Who will volunteer to do so?" (214).

According to Elof Axel Carlson, Muller valued the altruistic virtue that he called "love of one's fellows"; he believed that it was a superior genetic trait that the "best" people carried.[4] Perhaps he thought that physicists were the type of people who could be motivated by an appeal to this inherent virtue. But it is clear from the lack of response that this appeal did not work. Physicists were not interested in working "for" biologists. In addition, biologists were not motivated to learn about physics after reading Muller's text because Muller had implied that they could rely on altruistic physicists to do the work for them. Though it is true that Muller's lack of authority as a "mere biologist" was part of the reason he did not draw the attention of his audience, another part was that his essay was written in a way that maintained distance between the two disciplines and seemed to relegate the physicist to the role of a lab technician who would assist the biologist in solving genetic problems.

Niels Bohr and Max Delbrück

The temptation to view the benefits of collaboration only from the perspective of one's own discipline overcame the geneticist H. J. Muller. The physicists Niels Bohr and Max Delbrück must have faced a similar temptation. By arguing that biology can be fully reduced to physical explanation, they could have tried to authorize the physicist's seizure of the biologist's territory. Unlike Muller, however, Bohr and Delbrück were extremely careful to ensure that their discipline would not be perceived as sitting in a position of privilege over its counterpart in any interdisciplinary endeavor. Rather than argue for reductionism, they argued that biology could *not* be completely reduced to physical explanation.

This approach would seem to be a more promising start for interdisciplinary collaboration. It has the potential to negotiate the interests of the two disciplines rather than reducing one while privileging the other. But it did not spark the excitement that Schrödinger's did. I believe that Bohr and Delbrück ultimately failed for the same reason that Muller failed; their arguments maintained the distance between the two disciplines and

4. Carlson, "H. J. Muller," 777.

limited the perceived power of physicists who decided to engage in interdisciplinary activity.

Niels Bohr's appeal for collaboration came in his 1932 lecture (published in the journal *Nature* in 1933), titled "Light and Life." In this address, he proposed that "physics may influence our views as regards the position of living organisms within the general edifice of natural science."[5] As far as I know, Max Delbrück was the only individual who said he was inspired by this lecture to turn from physics to the study of biology.[6]

Delbrück's own appeal for collaboration was made in a 1949 lecture titled "A Physicist Looks at Biology."[7] Here, he argued that physicists who keep an open mind toward collaboration "will create a new intellectual approach to biology."[8] For the most part, Delbrück's lecture and the published essay that came from it went unheeded.

The most striking thing about both essays is that rather than adopt the reductionist approach of assuming that biology could be fully explained in terms of physical laws, they argued that some aspect of biological matter would ultimately remain hidden from the physicist's methods. Said Bohr,

> In every experiment on living organisms there must remain some uncertainty as regards the physical conditions to which they are subjected, and the idea suggests itself that the minimal freedom we must allow the organism will be just large enough to permit it, so to say, to hide its ultimate secrets from us. . . . Indeed, the essential nonanalyzability of atomic stability in mechanical terms presents a close analogy to the impossibility of a physical or chemical explanation of the peculiar functions characteristic of life.[9]

Delbrück shared Bohr's respect for the boundaries of the biological sciences, arguing that it would be arrogant to believe that life could be wholly reduced to physical explanation. He warned that the physicist should be prepared to find that life has placed an essential bar against its own reduction to the terms of molecular physics.[10]

By respecting the uniqueness of the biological sciences, Bohr and Delbrück honored the pride of biologists; they assured their readers that physicists were not seeking to colonize territory that rightfully belonged to another group of scientists. But this recognition of boundary lines also worked to maintain the distance between the two disciplines. At the same time that these authors promoted collaboration, they were implying that collaboration would break down at some level because the two disciplines

5. Bohr, "Light and Life," 312.
6. Olby, *Path to the Double Helix*, 231.
7. Delbrück, "A Physicist Looks at Biology."
8. Ibid., 22.
9. Bohr, "Light and Life," 317.
10. Delbrück, "A Physicist Looks at Biology," 17.

were too different to fully share the same rules. In addition, although Bohr and Delbrück did not explicitly minimize the professional status of physicists as Muller did, they did placc limits on the intellectual power of the physicist. Since physicists were the ones being asked to make the greatest change in professional goals, it is hardly an effective strategy to tell them that they would be unable to ever fully understand their subject matter. Even Delbrück admits that the irreducibility of biology is a "depressing" idea to physicists who may have thought that biology was an interesting field to enter.[11]

Schrödinger

What was really needed to inspire collaborative action was a negotiation of the two fields that did not obviously privilege one discipline over the other, nor limit the power of scientists who chose to embark on the new path of inquiry. This negotiation of interests was the contribution that Schrödinger's book made to the development of molecular biology. At the outset Schrödinger explained that his intention was "to make clear the fundamental idea, which hovers between biology and physics, to both the physicist and the biologist," namely, that physics and chemistry can account for events that take place in the living organism (*SW*, 1). Rather than ask the physicist to help the geneticist, and rather than argue that physics will never be able to fully account for biological events, Schrödinger addressed both groups with a claim that collaboration would yield scientific results. Also, as I argue later in this chapter, Schrödinger's ability to be ambiguous about the nature of those scientific results allowed readers with different philosophical and professional commitments to interpret his appeals differently, and thus unite in praise of the text.

A look at the responses of his readers indicates that Schrödinger's book was seen either as carefully negotiating the two fields or as supporting the particular disciplinary interest of a given reader. The book was not read by biologists as an attempt to wrongfully encroach on their territory, nor as an assurance that physicists would do their work for them. Neither was it read by physicists as a request that they volunteer their time to advance the careers of biologists, nor as a plea that they give up their own promising work for an ultimately fruitless study.

Among the biologists who wrote book reviews, H. J. Muller immediately recognized that Schrödinger's goal was similar to his own.[12] He also predicted that, unlike his own attempt, Schrödinger's interdisciplinary negotiation would be a successful one: "This little volume should prove valuable in furthering the much needed liaison between the fundamentals of the physical and the biological sciences. . . . It is to be expected that its function will mainly be to serve as a stimulus for inducing further progress

11. Ibid., 22.

12. He cites his own earlier article in the first paragraph of the review: Muller, "A Physicist Stands Amazed," 90.

in this direction."[13] The geneticist C. D. Darlington read the book as a bipartite discussion of "the progress of genetics and cytology on the one hand, and of physics and organic chemistry on the other." That progress brings a physical understanding of biology that "is both physical and biological in its importance."[14] David Allardice Webb, a botanist living in Dublin, also argued that the book was a merger of two fields. "The author's aim is ambitious; it is to set forth in ninety small pages the principles of quantum mechanics and of genetics, and to suggest a speculative connection between the two."[15] In addition, the naturalist Carl L. Hubbs and an anonymous reviewer for the medical journal *Lancet* both accepted Schrödinger's apology for an "act of seeming trespass" on the grounds that he was embarking on a much needed and quite successful "synthesis of facts and theories" between the two fields.[16]

Other biologists read the book as a description of how a study of physics could help biologists answer fundamental questions in their own field. For example, the geneticist J. B. S. Haldane was skeptical about the impact of the book on physicists, who, he believed, were much too interested in atomic energy to turn to the study of biology. On the other hand, said Haldane, "every geneticist will be interested in Schrödinger's approach to his or her science."[17] Likewise, the immunologist Peter Medawar saw the book's emphasis on the physics of the living cell as a "lively and interesting" addition to the study of biology, something that "is just what the professional biologist wants."[18]

Physicists who reviewed the book tended to read it as a evenhanded negotiation of disciplinary interests. According to the theoretical physicist Leopold Infeld, the book was not meant as "an answer to unanswerable questions" in the biological sciences. "It deals vividly and intelligently, however, with questions that lie on the borderline between physics and biology." In doing so, it introduces a new "field which many people believe will form the center of future research."[19] The physical chemist Michael Polanyi also predicted that the book would spark an interdisciplinary collaborative effort.

> [Schrödinger's book] will exercise its guiding influence on aspirants to a scientific career looking for new worlds to conquer. The sense of belonging together will be truly strengthened between biologists on the one side and physicists and chemists on the other. The book

13. Ibid.
14. Darlington, review of *What Is Life?* 126.
15. Webb, "What Is Life?" n.p.
16. Hubbs, review of *What Is Life?*, 554–55; "What Are We," n.p.
17. Haldane, "A Physicist Looks at Genetics," 375.
18. Medawar, review of *What Is Life?* n.p.
19. Infeld, "Visit to Dublin," 13.

will stimulate meetings and discussions between the two groups of scientists which may bear precious fruit.[20]

These excerpts show that though some scientists had slightly different views of the purpose of the book, they *all* saw it as an attempt to add something important to their own field of research, whether through a collaborative alliance or the importation of ideas and information. In the remainder of this chapter, I examine the textual devices that contributed to these positive readings of the book and its purpose.

The four rhetorical strategies that I discuss worked by directing the attention of distinct interpretive communities in different ways. In so doing, these strategies created in each reader's mind the most favorable possibility for a physical study of biology. The first strategy appealed to common values even as it spoke separately to physicists and biologists. The second appealed to the common goals of the two disciplines through two parallel but complementary statements about promising new areas of research. The third shifted the traditional linguistic practices of the two disciplines to subtly resituate professional loyalties and thereby promote cross-pollination of the fields. And the fourth strategy appealed to individuals with different ideological beliefs by promoting different readings of the same passage. All four rhetorical techniques worked to direct different readers, for different reasons, along the same path of research.

Negotiating Common Ground: The Value of Precision

The first thing needed in a book seeking an interdisciplinary response was an assurance that biology and physics share enough common ground to make collaboration possible. This need was especially evident in the first three chapters of Schrödinger's book, where he attempted to give both physicists and biologists some basic knowledge about what is normally studied on the other side of the disciplinary fence. The situation called for Schrödinger to divide the two disciplines, speaking to one and then the other about the fundamental theories of their counterparts. But such division runs the risk of further polarizing the two groups. What is interesting is the way Schrödinger maintained common ground despite the constraints of his task. Sensitive to the dual audience, he was able to emphasize similarities even while describing the differences between the two fields.

He began by speaking mainly to the biologist, explaining in chapter 1 that the laws of physics and chemistry are "statistical throughout" (*SW*, 2). In these pages he tried "to appeal to a reader who is learning for the first time about this condition of things," explaining to the biologist that this information is as fundamental to physics "as, say, the fact that organisms are composed of cells is in biology" (*SW*, 9). In contrast, chapters 2

20. Polanyi, "Science and Life," 3.

and 3 introduced the physicist to the basics of biology and genetics, explaining "the hereditary mechanism" and "mutations." Concentrating only on the essentials at this point, he apologized to the biologist half of his audience for "the dilettante character of my summary" (*SW*, 19).

This apology is just one of several that were placed throughout this section of the text. In most cases, Schrödinger made general statements then immediately retracted them in the footnotes for the benefit of those who possessed more sophisticated knowledge and who might have objected to the oversimplification. For example, in the text he wrote: "The actual size of atoms. . . ," while in the footnote he conceded: "an atom has no sharp boundary, so that 'size' of an atom is not a very well-defined conception" (*SW*, 5). With the text he proclaimed: "In meiosis the double chromosome set of the parent cell simply separates into two single sets"; with the footnote he admitted: "in fact, meiosis is not one division." (*SW*, 23–24). Accompanying these frequent retractions were apologies for the oversimplification. For instance, "The biologist will forgive me for disregarding in this brief summary the exceptional case of mosaics," and "I have the bad conscience of one who perpetuates a convenient error. The true story is much more complicated." (*SW*, 22, 49).

Although the first chapter was devoted to educating the biologist and the second and third were devoted to educating the physicist, there was no specific preview directing the two audiences to separate chapters, nor did Schrödinger ask either group to "skip ahead" in its reading. Schrödinger was addressing both audiences with these retractions and apologies; both could read them as a deference to precision. As a result, two groups would be reassured concerning the fundamental similarity between their disciplines, the common ground of precise science, while being made aware that there was much to learn about the complexities of the other field.

Negotiating Professional Goals: The Appeal to Ambition

Though all scientists share certain professional *values*, such as precision, the specific professional *goals* of biologists and physicists differ. As Delbrück explained in his 1949 lecture, the physicist seeks absolute laws about inanimate matter while the biologist seeks general regularities about living things.[21] An appeal for mutual action from the two groups would have to somehow transcend this difference. Rather than ask physicists to help the biologists reach their goals (as Muller did) or demand that biologists abandon their goals to study the absolute laws of physics, a successful interdisciplinary text would have to negotiate the professional goals of the two groups. One way in which Schrödinger's text was able to do this was by generating appeals from the general ambitions that all scientists share. The thrill of being recognized by one's peers and the promise of gaining

21. Delbrück, "A Physicist Looks at Biology," 9–10.

power over the natural world were professional objectives that both physicists and biologists could appreciate; so to inspire an interdisciplinary alliance, Schrödinger's text used an appeal to scientific ambition.

This appeal to fame is similar to Dobzhansky's inspirational calls. Using his ethos as a renowned scientist, Schrödinger advised the ambitious scientist reading his book that the best way to become successful would be to locate a project that is timely and promising. Focusing on the history of genetics, he pointed out that Mendel's discovery in the 1860s went unnoticed because the world was not yet ready for it. "Nobody seems to have been particularly interested in the abbot's hobby, and nobody, certainly, had the faintest idea that his discovery would in the twentieth century become the lodestar of an entirely new branch of science, easily the most interesting of our days" (*SW*, 41). The fact that the physicist Max Planck discovered quantum theory in the same year that Mendel's work was rediscovered is significant, said Schrödinger. Arguing that both genetic theory and quantum theory took time to mature before they could be productively connected, he proclaimed that now is the time when a "connection could emerge" (*SW*, 48). Later in the text, the same language used to characterize genetics appeared again, this time to evaluate the work of researchers who combine genetic mutation and atomic physics. "These facts are easily the most interesting that science has revealed in our day" (*SW*, 77).

There are other moments when the text tries to assure the audience that a physical study of the living cell holds promise for the scientist who engages it. For example, Schrödinger called the X-ray mutation studies of Timofféef-Ressovsky "beautiful work" that is "ingenious" in its application of physics to biology (*SW*, 42–43). He praised the recent success of "the united efforts of biologists and physicists" to determine the size of the gene (*SW*, 46). And he applauded the genetic work of the "German physicist" Max Delbrück (*SW*, 56). But these evaluations only served as a backdrop to the more grand proclamation that genetics is the branch of science that is "easily the most interesting of our days," and the subsequent proclamation that the physics of the genetic mechanism produces facts that are "easily the most interesting that science has revealed in our day." A young scientist seeking a productive field of research could hardly ask for more pointed advice than these approving words of wisdom from a famous Nobel Prize winner.

The ethos that Schrödinger carried was a significant component of this appeal to ambition; the same words offered by an unknown voice would hold little persuasive force. However, the way in which the appeal was worded as a double proclamation also contributed to its persuasive impact. Had Schrödinger merely remarked that genetics was the science "most interesting of our days" he may have alienated physicists who saw nothing of their own science in the proclamation. On the other hand, had he merely argued that the physical study of genes produced facts that were the "most interesting of our days," some biologists may have interpreted this statement as the reductionist move of a physicist who wanted to encroach on their disciplinary territory. The fact that Schrödinger made *both* remarks was significant to the inspirational impact of his appeal. For geneticists, the positive assessment of their field by a famous physicist could only serve to confirm their belief in the accuracy of his further claims about the timeliness of collaboration; their egos had been stoked by the first claim and their ambition was directed by the second. For physicists, Schrödinger's first claim was combined with the second claim to suggest that the genetic material was ripe for study by physicists who wanted to make significant new discoveries.

Although these two statements were deeply embedded in the text, rather than prominently displayed at the beginning or ending of a chapter, a look at the response of Schrödinger's readers suggests that this appeal to ambition was striking to the memories of both biologists and physicists. The geneticists Darlington, Muller, and Haldane all quoted Schrödinger's assessment of genetics in their book reviews.[22] Pleased that Schrödinger had proclaimed the superiority of genetics, all three of them went on to praise his attempt to inaugurate a physical study of genetics.[23] On the other side of the disciplinary fence, the physical chemist Polanyi was impressed by the claim that genetics was the most interesting branch of science, and he was additionally impressed by subsequent arguments about the potential success of collaboration. In his review he quoted Schrödinger's assessment of genetics right before predicting that young scientists seeking new worlds to conquer would be influenced by Schrödinger's book.[24] From the response of his readers, it appears that Schrödinger's use of parallel wording in these two inspirational calls positively influenced both the biologist and physicist contingents of his audience.

Negotiating Disciplinary Linguistic Practices: Conceptual Chiasmus

Just as the specific professional goals of the two disciplines differed, so too did the linguistic practices. Since physics traditionally works to understand the laws that operate on "inanimate" matter, the language of physics speaks mainly of objects that lack purpose, objects being pushed about by mathematical laws. On the other hand, the language of biology in Schrödinger's time was still occasionally tinged with animation, purpose, and the gleam of soul. Although the notion of vaporous spirits that animate the body had long since been abandoned, the vitalist force remained stubbornly at the center of the vocabulary used to describe cellular life.

22. Darlington, review of *What Is Life?* 126; Muller, "A Physicist Stands Amazed," 90; Haldane, "A Physicist Looks at Genetics," 375.

23. In fact, Darlington was so impressed with these statements that he actually combined them, slightly misquoting Schrödinger as saying that the new branch of science built around the physical study of the gene is "easily the most interesting of our day."

24. Polanyi, "Science and Life," 3.

Even strict reductionists such as Muller, who believed that genetics could be fully explained in terms of physical forces, relied on vitalist language to describe the action of the gene. He did so because it was the natural language of his field. For example, in his article urging physicists to contribute their time to the biologist's quest, he described the property of genetic "auto-attraction" as a mysterious interpenetrative force.

> Unlike the ordinary forces of adsorption known to the physical chemists, these gene forces are of such range as to act over visible microscopic distances. In doing so, moreover, they must in some way interpenetrate one another in many directions, since the forces of attraction of many genes must be traversing the same space at the same time. And despite this interpenetration, these forces must somehow preserve their directions and their specificities.[25]

To the skeptical reader, this force sounds suspiciously like that old vitalist bogy, "action at a distance." Likewise, vitalist language polluted Muller's description of a "second property of genetics."

> This second peculiar property is that of *auto-synthesis.* That is, each gene, reacting with the complicated surrounding material enveloping all the genes in common, exerts such a selectively organizing effect upon this material as to cause the synthesis, next to itself, of another molecular or supermolecular structure, quite identical in composition with the given gene itself . . . the gene actually initiates just such reactions as are required to form precisely another gene just like itself; it is an active arranger of material and arranges the latter after its own pattern.[26]

As this passage shows, to explain the property of auto-synthesis, Muller anthropomorphized the gene, giving it the ability to make intelligent choices and determine its own fate. The unequal power relations implied in Muller's request that physicists work on these problems for geneticists probably did much to attenuate the effect of his essay, but the vitalist taint in the explanations of these problems may have contributed to the failure as well. Most physicists were comfortable working with inanimate matter, not with active arrangers.

In a book that successfully connected physics and biology, the conflicting linguistic practices of these disciplines would have to be negotiated. The solution that Schrödinger stumbled upon was a reversal of linguistic practices. To make biology intelligible to the physicist, Schrödinger had to rid it of its mysterious vitalist presumptions, describing biology in clear mechanistic language. To make physics intelligible to the biologist, he had to describe the objects and concepts of physics in the language of purpose and action that was so familiar to the biologist.

25. Muller, "Physics," 211.
26. Ibid., 211–12.

He used this reversal strategy throughout the book; living matter was consistently described in inanimate physical language while inanimate matter was described with living, vitalist force. This reversal probably served two functions. First, when Schrödinger was instructing physicists about biology, his use of physicalist language made biological matter seem more accessible to them, and likewise, biologists found his description of physics accessible through the metaphors of life applied to it. Second, the reversal worked a sort of conceptual chiasmus, forcing each group to not only recognize their own study in the territory of the other discipline but to see their own territory opened to the conceptual tools of the other discipline. As biology became more mechanistic and physics became more vital, the conceptual apparatus of the two disciplines began to merge, making the possibility of meaningful collaboration more likely. Let us look more closely at how this was done.

In Schrödinger's text, atoms, the basic unit of physics, showed cooperative behavior, and heat motion was found "striving" against magnetic force in a rival "contest" (*SW*, 8, 10, 11). Permanganate molecules also exhibited purposive action. "Every one of them behaves quite independently of all the others, which it very seldom meets" (*SW*, 13). Atoms continue their vitalist agenda as they "play a dominating role in the very orderly and lawful events within a living organism. They have control . . . they determine" (*SW*, 19). But though atoms and physical forces were given vital energy, "bacteria and organisms" had this purposiveness taken away: "they have no choice" as heat motion "tosses" them around (*SW*, 11, 12). The human body was made mechanical when Schrödinger argued that "our organs of sense, after all, are a kind of instrument" (*SW*, 15). And what may be considered the most vital living forces of all—egg and sperm—merely "coalesce" in the author's description of fertilization (*SW*, 23).

Sometimes Schrödinger found it too difficult to make the switch. But when figurative language of agency slipped into a passage on biology, quotation marks undermined any suggestion of purposive action. With genetic mutation, "some change is 'tried out' in rather a haphazard fashion" (*SW*, 41). The author had no problem anthropomorphizing atoms but felt the need to underline the metaphorical nature of his description when it came to gene mutation. Likewise, Schrödinger used scare quotes to call into question the organism's purported vitalist ability to escape the second law of thermodynamics, while resituating that ability in the individual active atom.

> An organism's astonishing gift of concentrating a "stream of order" on itself and thus escaping the decay into atomic chaos—of "drinking orderliness" from a suitable environment—seems to be connected with the presence of the "aperiodic solids," the chromosome molecules, which doubtless represent the highest degree of well-ordered atomic association we know of—much higher than the

> ordinary periodic crystal—in virtue of the individual role every atom and every radical is playing here. (*SW*, 77)

The astonishing gift of an organism's agency was given an ironic treatment through the scare quotes, while the role playing of the atom was presented as a straightforward account of its power over the motions of the organism.

Likewise, in other parts of the text, quotation marks were employed to counter any suggestion that mechanistic language could be used to describe the activity of atoms:

> A number of atomic nuclei, including their bodyguards of electrons, when they find themselves close to each other, forming "a system," are unable by their very nature to adopt any arbitrary configuration we might think of. Their very nature leaves them only a very numerous but discrete series of "states" to choose from. (*SW*, 49)

Since "system" and "states" are more standard physics fare, suggesting impersonality, they could only be used with caution, their not-quite-right nature spotlighted by the quotation marks. Although a level of determinism was suggested by the sense of the paragraph, the atoms were left with the qualities of choice and social organization that offset any suggestion of mechanism.

Schrödinger's rhetorical reversal of conventional usage is made clear by the inconsistent use of such quotation marks, for the word *system* appeared later without these marks. "To the physicist—but only to him—I could hope to make my view clearer by saying: The living organism seems to be a macroscopic system which in part of its behavior approaches to that purely mechanical . . . conduct to which all systems tend" (*SW*, 69–70). This time, the system-machine metaphor was connected to a living organism, not an atom or physical force, so the association could be entertained without hesitation.[27]

Schrödinger made it clear that he would like only *physicists* to think about the mechanical, systematic activity of the organism, so one would presume that he would have biologists focus their attention on the vitalist activity of the atom. Perhaps he wanted biologists to look toward the metaphor of magic he used to describe the atoms and molecules of physics

27. The rhetorical critic Carolyn Miller suggested to me that there is another way of reading Schrödinger's use of scare quotes. Since Schrödinger was talking mainly to physicists when he discussed biology, and vice versa, it is possible that the scare quotes are his way of saying "this is how biologists talk (although we physicists wouldn't say it that way)" and vice versa. This would make each group feel that Schrödinger was savvy to that group's way of thinking, and would create a kind of camaraderie within each group by negating the linguistic practices of the other. Her reading makes sense. While the scare quotes maintained the consistency of his linguistic reversal, and thus ultimately helped reduce the conceptual distance between the two disciplines, they also may have increased distance between the two groups by belittling the linguistic practices of the other. Carolyn Miller, letter to author, 5 January 1997.

but not the living organism. A large molecule is "a masterpiece of highly differentiated order, safeguarded by the conjuring rod of quantum theory" (*SW*, 69). A small group of atoms "displays a most regular and lawful activity—with a durability or permanence that borders upon the miraculous" (*SW*, 46). The quantum jump "is a rather mysterious event" (*SW*, 49). And while these mystical atomic forces substitute for the vitalist energies of living matter, life is found to be pure mechanism. Biological functions are metaphorically described as "the working of a large manufacturing plant in a factory" (*SW*, 41).

In playing on the metaphorical presuppositions of the two disciplinary vocabularies, Schrödinger was able to appeal to each side in the language that was most familiar to it, while getting each side to think of its opposite in the terminology most often applied to itself. Physicist and biologist were left believing that what they understood in their chosen field of study could be found in greater abundance on the other side of the disciplinary fence—some physicists in the audience found uncharted territory in the mechanism of the living body, while some biologists in the audience found the key to the vitality of life in the physicist's atom.

In his review of *What Is Life?* Muller objected to the fact that Schrödinger did not discuss the genetic property of autosynthesis. But perhaps the reason Schrödinger chose not to address this property is that it would have required him to emphasize a vitalist activity of the gene and fall into the standard language patterns that had for so long served to separate the two disciplines. In the same review, Muller recognized that Schrödinger's use of language, though not always the most traditional, "has, however, the value, for his purposes, that it may serve better, for readers of the type to whom he is chiefly addressing himself."[28]

Of course, the effectiveness of Schrödinger's linguistic reversal strategy does not depend on its being consciously recognized and adopted by the readers. All it had to do was make the readers feel uncomfortable about locating a vitalist spirit in the organism and more comfortable about physicalizing the gene. For example, the immunologist Peter B. Medawar did not adopt Schrödinger's strategy when he summarized parts of the text in his book review. In fact, Medawar consistently shifted the linguistic practices back to their traditional forms. Describing Schrödinger's writing, Medawar explained that "organisms as a whole behave in a regular and orderly fashion," while the "system of atoms" that make up a gene is "knit together . . . by reason of [the atoms'] existing in the lowest energy level." In this description, agency goes to the larger biological units and is taken away from the atoms. But later in the review, Medawar showed that although he had not been persuaded to reverse the linguistic practices

28. Although Muller's compliment was not specifically referring to the linguistic shift that vitalized the atom and physicalized the gene, I believe he was right to highlight the effectiveness of Schrödinger's vocabulary choices. Muller, "A Physicist Stands Amazed," 90.

himself, he had been persuaded to look askance at the use of vitalist metaphorical language by others to describe biological objects. In his review, Medawar objected to a case in which Schrödinger described an organism in vitalistic terms. (In this case Schrödinger applied scare quotes to highlight the *inadequacy* of vitalism when applied to organisms, but Medawar did not make this distinction when critiquing the metaphor.) Referring to the passage cited earlier, in which an organism is said to be "drinking orderliness from the environment," Medawar complained about the metaphorical language Schrödinger used.[29] Medawar did not recognize the irony in Schrödinger's use of this metaphor, and so he claimed that Schrödinger was making poor language choices. But it is at least possible that the reason Medawar objected when he saw this vitalistic language applied to the organism was that he had been subconsciously influenced by Schrödinger's reversal strategy! Though it was still natural for this biologist to talk about biological organisms as agents with vital purpose (as his own language choices demonstrated), he had begun to consider that language, when used by others, to be unilluminating. Presumably, the use of language that was less metaphorical and more concrete, that treated the biological organism as a thing rather than an active agent, would have been more illuminating to this biologist, who was beginning to think of the living organism in physical terms.

Negotiating Ideological Commitments: Strategic Ambiguity

The final rhetorical strategy is perhaps the most interesting. As outlined in the previous chapter, Schrödinger's "other laws of physics" passage has been read by historians to mean three different things. In this chapter, I argue that the rhetorical construction of the passage and of the text that surrounds it invited all three readings from the book's original audience, and it was at least partially due to this ambiguity that the book appealed to readers with very different ideological commitments.

To understand how the different reading frames appealed to different audience groups, we must first understand the historical context in which Schrödinger was writing. Long before and after Schrödinger introduced the possibility of "other laws," there were countering currents of reductionism and antireductionism that guided individual researchers in both the biological and the physical sciences. Reductionists argued that all of science could ultimately be reduced to relatively simple deterministic physical laws; antireductionists argued that this was an overly simplistic hope that did not take into account the acausal character of some natural events.

In biology, the tradition of reductionism went as far back as Descartes "L'Homme Machine," which presented a mechanistic model for the human body.[30] But it was the late nineteenth and early twentieth centuries

29. Medawar, review of *What Is Life?* n.p.
30. C. U. M. Smith, *Problem of Life*, 159–76.

that saw the rapid rise of reductionism as a powerful philosophy in the biological sciences. In 1911, Jacques Loeb made it his mission to dispel vitalist explanations and prove that the wonders of reproduction were nothing more than the mechanistic interaction of chemicals that operate by simple physical laws. According to Loeb, "As long as a life phenomenon has not yet found a physico-chemical explanation it usually appears inexplicable. If the veil is once lifted we are always surprised that we did not guess from the first what was behind it."[31] Loeb was not the only one to take on the mission against vitalism; according to John Fuerst, reductionism was a critical belief system that guided many of the important scientists who would contribute to the development of molecular biology.[32] In the physical sciences, reductionism was also a strong philosophical current. For example, the physicist Ludwig Boltzmann, whom Schrödinger strongly admired, argued that the reason the atomistic theory of the materialists was a good explanation of physical matter was that it reduced the world to understandable mechanical models.[33]

But antireductionism was also a philosophy with a strong following in the twentieth century. In biology, Wolfgang Ostwald, the son of the idealist physicist Wilhelm Ostwald and a former student of Loeb, argued that classical physical chemistry was inadequate to explain the action of biological cells. He suggested that entirely new forces would have to be discovered to understand the working of the living cell.[34] Hans Driesch argued in 1935 that experiments with heredity proved the inadequacy of reductionist accounts and gave new life to an antimechanist, vitalist philosophy. "To put it shortly: organic life is *autonomous*, i.e., it is not a mere combination of single events of the inorganic type."[35] In physics, reductionism also suffered a great defeat in the twentieth century when the special theory of relativity pointed out the ultimate limitations of scientific observation and when quantum mechanics proved the acausal nature of atomic states. When Niels Bohr argued for the presence of complementary explanatory schemes for physical phenomena, he was just one of many to conclude that a simple reductionistic account was no longer possible for physics, much less for biology. As explained earlier in this chapter, Bohr and Delbrück both argued that a reductionist approach would ultimately fail to fully explain the secrets of life.

It was in this environment of long-lasting conflict between reductionism and antireductionism that Schrödinger's "other laws of physics" passage was born, and it was in this environment that it was read by both

31. Loeb, "Mechanistic Conception of Life," 28; see also Pauly, *Controlling Life*.
32. Fuerst, "Role of Reductionism," 269.
33. Boltzmann, "Necessity of Atomic Theories," 357–71.
34. Pauly, *Controlling Life*, 150–51.
35. Driesch, "Breakdown of Materialism," 290.

physicists and biologists. Unlike the other rhetorical strategies in Schrödinger's book, this one did not appeal separately to the professional goals or linguistic practices of the two disciplines. Instead, this passage appealed to ideological commitments that cut across disciplinary boundaries; some physicists and biologists read the passage as antireductionist, while others read it as reductionist.

Recall the three reading frames that modern historians have offered: Stent argued that Schrödinger supported the idea of acausal "new laws" of physics hiding in the living organism; Olby argued that Schrödinger was talking about new "order-from-order" laws that would recover determinism in the physical sciences; and Fleming argued that Schrödinger was referring to the "Heitler-London forces." In fact, all three of these reading frames were used by Schrödinger's original audience: an interpretation from Stent's frame appealed to antireductionist scientists, while readings from Olby's and Fleming's frames appealed to reductionists.

Stent's Reading Frame

A reading of "other laws" as complementarity laws that have nothing to do with current physical principles comes readily from the famous passage itself, which is written in a way that lends itself to this sort of direct interpretation. If the laws are "other laws of physics" that were "hitherto unknown," they would seem to be "new" laws that differ in some way from the current ones. What would be easier than to assume that they differ because they retain something of the mysterious, nondeterministic nature of traditional explanatory schemes in the biological sciences? Readers were set up for this interpretation as early as the first two pages of the book, where Schrödinger claimed that "present-day physics and chemistry could not possibly account for what happens in space and time within a living organism" (*SW*, 2). Later, Schrödinger made a nod to Bohr's complementarity theory when, at two different points in the text, he recognized the ultimate irreducibility of organic events such as "human thought" (*SW*, 7, 31).

The strongest textual support for Stent's reading frame came when Schrödinger explicitly compared a mutation to a quantum jump. Describing de Vries's mutation theory, he said: "The significant fact is the discontinuity. It reminds a physicist of quantum theory—no intermediate energies occurring between two neighbouring energy levels. He would be inclined to call de Vries's mutation theory, figuratively, the quantum theory of biology" (*SW*, 34). After reading this explicit metaphor, the audience would associate the tenor (mutation theory in biology) with the vehicle of those discontinuous, nondeterministic, somewhat mysterious laws of quantum theory. Making this association, the audience might come to believe that like quantum events, mutations are an exciting new discovery in biology, and that just as physicists recently discovered the new "quantum theory" that guides quantum events, so too would physicists soon discover the new quantum laws that guide mutation.

What was especially inspirational about this reading of the "other laws of physics" passage was that it was antireductionist (maintaining that biology could not be reduced to simple deterministic laws of physics) without putting a limit on the ambition of physicists who decided to take up the challenge (as the antireductionist messages of Bohr and Delbrück had done by admitting that the uniqueness of life would allow the organism to "hide its ultimate secrets" from the physicist). The physicist who was not opposed to the idea of nondeterministic laws would have found an exciting challenge in the hunt for new laws of physics that would unravel the secret of life. At the end of the book, after "reviewing the biological situation," Schrödinger concluded that "the point to emphasize again and again is that to the physicist the state of affairs is not only plausible but most exciting, because it is unprecedented" (*SW*, 77–78). Enterprising young physicists reading the book would understand this state of affairs to be one in which a physicist need only turn to the "virgin territories" of biology to discover a new physical principle, a discovery that would most certainly secure the highest level of scientific fame.

Stent gave much weight to the psychological influence of this reading, claiming that it motivated him and many others who were to work on the physics of biology. But Stent's claim that others were inspired was speculative, and he was forced to qualify it greatly after asking four fellow molecular biologists to recall their interpretation of the passage.[36] On the other hand, a look at intertextual evidence of responses written soon after Schrödinger's book was published suggests that Stent was not the only physicist to be captivated by this idea. In 1947, two engineers took this reading to heart when they published a theoretical study on "quasi-properties," which they defined as fundamentally different from normal "physical properties." In this article they cited Schrödinger's book as one that postulates a physical difference between living and nonliving matter.[37] In 1950, two theoretical physicists wrote an article titled "Living Matter and Physical Laws," which they said was "considerably stimulated by reading Schroedinger's book."[38] In it they spoke about the search for a "new 'life' principle," a physical law for biological associations that are too complicated to be understood by a "straightforward application of quantum mechanics": "There are no methods known to physics at the present time for calculating these intermediate-sized aggregates [the enzymes]; but one might conjecture that if such a method were discovered, it would allow great advances in biology."[39] Although the individuals who

36. Stent, "Molecular Biology," 395.
37. Blair and Veinoglou, "Limitations of the Newtonian Time Scale," 69–87.
38. Reitz and Longmire, "Living Matter," 15.
39. Ibid., 15, 18.

wrote these words were not superstars of the molecular biology revolution, they were working scientists who were motivated to study the borderline between physics and biology because they thought that Schrödinger had predicted that revolutionary "other laws" of physics would soon be found there. Had Stent asked more than four people, had he asked them in the years immediately following the publication of Schrödinger's book (before the hope for new laws was revealed to be an illusion and people began to forget their earlier feelings about it), and had he cast his net wider than the "core group" of famous molecular biologists, he might have found more support for his thesis that physicists were influenced by this reading frame.

For biologists, Stent's reading frame may also have held some appeal. In the more recent historical accounts, it is the biologist-turned-historian (for example, François Jacob or Ernst Mayr) who offers the most unhesitating support for this reading. I think the reason for this is that Stent's nonreductionist reading maintains a distinct place for biology in the future of science. In Schrödinger's time, biologists who were afraid that the more powerful physicists were trying to reduce their science to a subset of physics may have read the call for new nondeterministic laws as an assurance that biology would retain a unique character of its own. For example, one reviewer in the medical journal *Lancet* in 1945 was sensitive to the possibility that a book such as Schrödinger's would turn out to be the inappropriate "irruption of a physicist into the realms of biology," but after reading the book, he recognized the antireductionist stance of the author. According to this biologist, Schrödinger's book argues that "living matter may evade the laws of chemistry and physics," and this means that we must continue to search for a synthetic picture of life. Such an antireductionist reading made *What Is Life?* "an exciting, thought-provoking book" for this particular biologist, who saw Schrödinger's monograph as a successful synthesis of biology and physics, rather than an attempt at a hostile takeover.[40]

In short, for some physicists *and* for some biologists, the idea of a nonreductionist search for "new laws" in the biological organism was a conceptual commitment that fit well with their professional goals and disciplinary concerns. Had the text not been written in a way that allowed these individuals to interpret it this way, they may not have responded so favorably to Schrödinger's call for collaboration between the fields.

Olby's Reading Frame

Olby's interpretation is also plausible as long as the reader approaches the text with a certain predisposition. Although the idea of Bohr's philosophy can be attached to the text by a reader who is looking for it, Schrödinger never explicitly invoked the idea of complementarity, and in the

40. "What Are We," n.p.

epilogue, he seemed to actually attack Bohr's position when he said that "in my opinion, and contrary to the opinion upheld in some quarters, *quantum indeterminacy* plays no biologically relevant role" in the activity of the organism (*SW*, 88). Of course, if the epilogue is read last, it does nothing to direct reading patterns throughout the book. In the preface, however, there is another passage that sets the stage for a reductionist reading of Schrödinger's text. In the second paragraph of the preface, Schrödinger confessed that he was writing out of a "keen longing for unified, all-embracing knowledge" (*SW*, vii). For the reductionist-minded, such a quest could mean the attempt to overcome the acausality of quantum physics with new deterministic laws, laws that truly explain the universe, rather than further mystify it with complementary explanatory schemes.

Although Schrödinger started his first chapter with the claim that present-day physics and chemistry cannot account for what happens in the living organism, he accompanied this claim with the promise that this inability "is no reason at all for doubting that they can be accounted for by those sciences" (*SW*, 2). This suggests a reductionist project similar to Loeb's, a project that assumes that our current lack of a physico-chemical explanation is due merely to the fact that the veil has not yet been lifted. Later in the book, when Schrödinger offered his motor engine metaphor, he seemed to directly call on this reductionist vision. In a section titled "New laws to be expected in the organism," he described those new laws as follows:

> What I wish to make clear in this last chapter is, in short, that from all we have learnt about the structure of living matter, we must be prepared to find it working in a manner that cannot be reduced to the ordinary laws of physics. And that not on the ground that there is any "new force" or what not, directing the behaviour of the single atoms within a living organism, but because the construction is different from anything we have yet tested in the physical laboratory. To put it crudely, an engineer, familiar with heat engines only, will, after inspecting the construction of an electric motor, be prepared to find it working along principles which he does not yet understand. He finds the copper familiar to him in kettles used here in the form of long, long wires wound in coils; the iron familiar to him in levers and bars and steam cylinders is here filling the interior of those coils of copper wire. He will be convinced that it is the same copper and the same iron, subject to the same laws of Nature, and he is right in that. The difference in construction is enough to prepare him for an entirely different way of functioning. He will not suspect that an electric motor is driven by a ghost because it is set spinning by the turn of a switch, without boiler and steam. (*SW*, 76).

In this passage Schrödinger told his readers that they should not be satisfied with the belief that the organism is driven by a mysterious new "life principle" (nor, one suspects, should readers be satisfied with the claim that the atom is driven by a mysterious quantum theory); instead, they should further study the phenomena for simple answers that are coherent with the classical laws of physics. A closer inspection of the special construction of biological matter will soon reveal the new laws that exorcise the ghost in the machine.

Schrödinger went on to describe these new laws as "order-from-order" laws that differ from the statistical laws of modern physics. In the study of inanimate matter, said Schrödinger, "we find complete irregularity, cooperating to produce regularity only on the average" (*SW*, 78). "In biology we are faced with an entirely different situation. A single group of atoms existing only in one copy produces orderly events, marvellously tuned in with each other and with the environment according to most subtle laws" (*SW*, 79). These subtle laws are unknown because the physicist and the chemist never took the time to investigate living matter. "The case did not arise and so our theory does not cover it—our beautiful statistical theory of which we were so justly proud because it allowed us to look behind the curtain" (*SW*, 80). Indeed, the beautiful theory becomes an ironic mistake as we find that the curtain was a false one. "It appears that there are two different 'mechanisms' by which orderly events can be produced: the 'statistical mechanism' which produces 'order from disorder' and the new one producing 'order from order.' To the unprejudiced mind the second principle appears to be much simpler, much more plausible. No doubt it is" (*SW*, 80). According to Schrödinger, because physicists spent so long focusing on the statistical laws, they neglected the more plausible ones. Physicists can correct that error by studying the biological organism, in which they will have to be "prepared to find a new type of physical law" (*SW*, 81).

It took little imagination for the reductionist-minded scientist who read these claims to recognize these new laws as hidden deterministic principles that would soon explain matter at a new level and thereby banish the unsatisfactory statistical laws of modern physics. The physicist who was unhappy with the current state of the discipline may have been inspired by this reading to study the living cell in search of these new deterministic laws. Thus, it is not surprising that the physicist-turned-molecular biologist Neville Symonds interpreted the passage through Olby's reading frame; after all, the influence that drew him to biology was the idea that it would be less paradoxical than modern physics, whose "every calculation . . . gave infinity as an answer."[41] For physicists seeking order in a universe gone mad, biology may have held out some hope as a place where rational reductionist laws could be discovered.

41. Symonds, "Schrödinger's Influence on Biology," 224. See also Symonds, "Schrödinger and *What Is Life?*" 663. I should note that Symonds said he was inspired to study biology by Schrödinger, but through conversations with him, not directly from his book. See "Schrödinger's Influence on Biology," 224.

Leopold Infeld was another physicist who adopted this reading frame. In his 1949 review of Schrödinger's book, Infeld quoted the famous "other laws" passage before going into a detailed description of the fantastic *order* that guides even the smallest organic processes.

> The mechanism that governs an organism which is composed of not too great a number of molecules is not the statistical probability mechanism of ordinary physics. We shall understand the laws of living organisms only when we understand the transition order-from-order which seems to reign in biology, just as the transition order-from-disorder reigns in physics.[42]

It is significant that Infeld described himself in that article as a theoretical physicist who worked with Einstein to derive stellar laws of motion from the general theory of relativity, laws that were as reduced and refined as they could possibly get.[43] For the scientist committed to a reductionist account of the natural world, and particularly for the physicist who was unsatisfied with the current direction that physics seemed to be taking, a reading through Olby's frame would have been the most effective motivator for interdisciplinary work.

Fleming's Reading Frame

Fleming's reading of the "other laws" passage is the most difficult to make from the isolated passage, but it is the reading that is most supported by the rest of the book. The famous passage is located at the end of the first section in chapter 6 of Schrödinger's book. At the beginning of the next section, the reader who is inclined to adopt Fleming's reading frame is given the first hint that a literal reading of the passage is incorrect. Said Schrödinger, "This is a rather subtle line of thought, open to misconception in more than one respect. All the remaining pages are concerned with making it clear" (*SW*, 69).

There are other clues embedded in the text to call the literal reading of Schrödinger's words into question. At two points in the book, footnotes foreshadow a shift in meaning. On the second page, where Schrödinger first described the laws of physics as "statistical laws," and again, in the section immediately following the "other laws" passage, where he repeated this description, Schrödinger inserted footnotes that retracted his "too general" statements and promised a fuller discussion in chapter 7. That chapter is titled "Is Life Based on the Laws of Physics?" and immediately following the title is an epigraph in Spanish, attributed to Miguel de Unamuno. Translated in a footnote, it reads: "If a man never contradicts

42. Infeld, "Visit to Dublin," 14.

43. Ibid., 11–12. Infeld did not enter the interdisciplinary space himself, but his interpretation represented how a physicist at the time would approach the book. In addition, his popularization of what I am calling the "Olby reading frame" in *Scientific American* may have influenced others to read Schrödinger's book in this way.

himself, the reason must be that he virtually never says anything at all" (*SW*, 76). With the idea of contradiction in mind, a reader might think that Schrödinger's earlier claim that life is *not* based on the current laws of physics was about to be retracted.

The first section of chapter 7 is titled "New laws to be expected in the organism," but it is not clear whether that is a declarative statement with the predicate understood or a noun clause that serves as title for the discussion that follows. Recall that this section is devoted to the electric motor analogy, an analogy that assures the reader that there is no new force driving the organism, only a difference of construction that the physicist may not yet comprehend. In the next few pages, Schrödinger made a remark that suggested he actually *did* comprehend that difference of construction.

Before getting to that remark, we must turn back to an earlier section of the book. In chapter 4, Schrödinger had hinted that the quantum mechanical Heitler-London forces, which hold molecules together, would suffice to explain the ordered stability of organic material. In those pages, he argued that a physicist at the end of the nineteenth century would have tried to understand the stability of the hereditary molecule, but would have been unable to do so because that stability is founded on the very basis of quantum theory (*SW*, 47). Schrödinger then went on to describe the Heitler-London forces (the stabilizing forces that work between atoms in a molecule) as the relevant aspect of quantum theory (*SW*, 50, 60). But it is not until the end of the book that these very forces reappear as the "new laws" that explain the remarkable order of the biological organism.

In a section near the end of the book titled "The new principle is not alien to physics," Schrödinger remarked: "[T]he new principle that is involved is a genuinely physical one: it is, in my opinion, nothing else than the principle of quantum theory over again. To explain this, we have to go to some length, including a refinement, not to say an amendment, of the assertion previously made, namely, that all physical laws are based on statistics" (*SW*, 81). In the next few paragraphs, Schrödinger explained that modern physics applies statistical laws to calculate the action of small numbers of atoms, but the classical laws of physics still apply to larger-scale events such as the motion of the planets and the action of "clockwork solids": "In short, all purely mechanical events seem to follow distinctly and directly the 'order-from-order' principle" (*SW*, 81). This amendment allowed Schrödinger to conclude:

> Thus it would appear that the "new principle," the order-from-order principle, to which we have pointed with great solemnity as being the real clue to the understanding of life, is not at all new to physics. . . . We seem to arrive at the ridiculous conclusion that the clue to the understanding of life is that it is based on a pure mechanism, a "clock work." . . . The conclusion is not ridiculous. . . . Clockworks are capable of functioning "dynamically," because they

> are built of solids, which are kept in shape by London-Heitler forces, strong enough to elude the disorderly tendency of heat motion at ordinary temperature. (*SW*, 82, 85)

With these words, we have reached the climax of the story Schrödinger was telling. The idea of "new laws" hiding in the organism is "contradicted" when he reveals the fact that London-Heitler forces explain the remarkable order of biological matter. All the clues led us along the trail to this final treasure, the understanding of life as pure mechanism, "not of coarse human make, but the finest masterpiece ever achieved along the lines of the Lord's quantum mechanics" (*SW*, 86).

For the reductionist-minded reader, an interpretation of Schrödinger's "other laws" passage that recognized them to be the London-Heitler forces would have been a powerful motivator for interdisciplinary study. It would have worked especially well on the biologist who was philosophically committed to a reductionist approach. With a reading from what I am calling "Fleming's frame," the biologist was assured that a study of the physical nature of biological matter would not require the discovery of strange new physical laws, acausal or causal; instead, the biologist would begin to find immediate results from the application of the physicist's methods to the organism.

In recalling his work under Max Delbrück, James Watson indicated why a biologist who wanted to make immediate contributions to biological questions might not want to interpret Schrödinger's book from what I am calling "Stent's frame": "Delbrück talked about Bohr and his belief that a complementarity principle, perhaps like that needed for understanding quantum mechanics, would be the key to the real understanding of biology. . . . So, sometimes I worried that my inability to think mathematically might mean I could never do anything important."[44] Fortunately, before he met Delbrück and was indoctrinated in the ineffable mysteries of the gene, Watson read Schrödinger's book and interpreted it as an assurance that a biologist who learned some physics would be able to penetrate the essence of life.[45]

Watson also had the benefit of working closely with Crick, a reductionist-minded physicist who believed that the main point of Schrödinger's book was to convey the idea "that biology needs the stability of chemical bonds and that only quantum mechanics can explain this."[46] Crick is committed to the belief that biology can be thoroughly reduced to physical mechanisms, and he has argued that many workers in the interdisciplinary field felt the same way; they were "strongly motivated by the desire, rarely actually expressed, to refute vitalism."[47] This desire would have been

44. J. D. Watson, "Growing Up," 240.
45. J. D. Watson, *Double Helix*.
46. Crick, "Recent Research," 184.
47. Ibid.

supported by a reading of Schrödinger's "other laws" passage from Fleming's frame.

Several biologists who wrote book reviews of *What Is Life?* seem to have interpreted the passage in the way that Watson and Crick did. The geneticist C. D. Darlington said that the book was written to show how the "prodigious orderliness of the chromosome molecule" has "enabled life to get away from the drawbacks of quantum indeterminacy. . . . In other words integration defeats indeterminacy."[48] He seemed particularly interested in this move, arguing that Schrödinger should continue the fight against vitalist biology: "Biology shows us integration passing from one level to another and reversing as it goes the appearance (or, as the reviewer would say, the *illusion* of indeterminacy) [*sic*]. . . . Professor Schrödinger has shown us the first steps on this way. We may hope that he will be willing very soon to conduct us further."[49] Another biologist who read the book in this way was an anonymous reviewer for the journal *Plant Breeding Abstracts*, who saw the book as "one of the most luminous accounts of the significance of physical theory to biology" and urged biologists to read it. This reviewer said that the "central thesis is that life is predominantly a manifestation of the orderliness of the solid state," and that the "constancy of the genotype" is explained "by quantum considerations."[50] Like Darlington, this reviewer was impressed by the fact that the specter of quantum indeterminacy had been ruled out through the support offered for the book's central thesis.

Yet another biologist who read the book through Fleming's frame also appeared to have loyalties to a reductionist philosophy. An anonymous reviewer for the *British Medical Journal* began the essay by expressing excitement for the fact that Schrödinger's book would soon allow us to solve the perennial problem of "making life in the laboratory." The mechanist dream of fully understanding the living in terms of the nonliving (and thus bridging the impassable gulf that keeps men from making life) was one that this reviewer saw actualized in Schrödinger's book. When describing the "other laws," this reviewer called them "different," not "new," pointing out that they are really just "the presence of aperiodic solids, the chromosome molecules in living matter."[51] This reviewer, like the other two, read the book as a reductionist answer to the question "What is life?" and, from such a reading, became excited to support Schrödinger's call for a physical study of the living cell.

The Persuasive Power of Polysemy

Just as with Dobzhansky's successful move toward synthesis, Schrödinger's relied heavily on techniques of polysemy. Bohr and Delbrück appealed

48. Darlington, review of *What Is Life?* 126.
49. Ibid, emphasis in original.
50. Review of *What Is Life?* in *Plant Breeding Abstracts*, 57.
51. Review of *What Is Life?* in *British Medical Journal*, n.p.

to only one type of reader with their call for complementarity and the concomitant claim that biology could never be completely reduced to physical explanation. In contrast, Schrödinger's call for action was more ambiguous; some read it as compatible with Bohr's thesis but more exciting because it promised new complementary laws of physics just waiting to be discovered, while others read it as a thoroughly reductionist account. As a result, Schrödinger's book appealed to a wide variety of readers, convincing each group that a move in the direction of "the physical study of the living cell" would be in their best interest. At least five different interpretive communities were addressed with this passage: physicists who were ready to start a quest for exciting new paradoxes in the virgin territories of biology; biologists who wanted to benefit from an alliance with physics but still retain some level of uniqueness for the living organism; physicists who were unhappy with the current paradoxes in physics and were interested in finding a new all-embracing determinism; biologists who wanted a quick and easy new tool for revealing the secrets of life; and physicists who believed that the current laws of physics would suffice to explain biological processes. All these groups, with their differing and even conflicting ideological and professional desires, were led by the ambiguity of the passage to believe that Schrödinger's call for a physical study of biology was one that spoke directly to their own interests (see table below).

Of course, it is important to note that as with Dobzhansky's text, polysemy does not always work in favor of the author's call for action. For example, in an article on the subject of organic development, the zoologist Sydney Smith quoted Schrödinger in order to dispute his alleged belief in new laws: "Since orderliness can be expressed in physical terms by invoking the concept of entropy (see Schrödinger, 1944), there has been a tendency in discussing the development of structural pattern in living things,

	Disciplinary Affiliation	
Philosophy	**Physics**	**Biology**
Antireductionism	*Stent's Reading Frame* Stent Blair and Veinoglou Reitz and Longmire	*Stent's Reading Frame* Jacob Mayr Reviewer in *Lancet*
Reductionism	*Olby's Reading Frame* Symonds Infeld *Fleming's Reading Frame* Crick	*Fleming's Reading Frame* Watson Darlington Reviewer in *Plant Breeding Abstracts* Reviewer in *British Medical Journal*

Reading frames adopted by different interpretive communities

to suppose that they evade in some way the limitations of physical laws."[52] This biologist read the book from Stent's frame, recognizing the call for "new laws" as a call for principles that are somehow extraphysical. But as a researcher who was perfectly happy to see the thermodynamics of living and nonliving matter run by the same rules, Smith was driven by that interpretation to counter Schrödinger's book. Likewise, the physicist L. Brillouin wrote a lengthy article that interpreted Schrödinger in a way that did not fully promote Brillouin's own ideological position. Brillouin supported the "revolutionary idea" that "life cannot be understood without reference to a 'life principle,' " but rather than read Schrödinger from Stent's frame and declare him a kindred spirit, he pegged Schrödinger as a more moderate thinker. Describing Schrödinger as the sort of person who can only admit that there *might* be some new physical law that, once discovered, will turn out to be rather mundane, he interpreted Schrödinger from Olby's frame and concluded that "Schrödinger's explanations are too clever to be completely convincing."[53]

Despite these two instances of readers interpreting the text against their own interests, polysemy seemed to work productively in most cases, and when it failed, it only promoted disputes of Schrödinger's arguments, not his mission to unite fields. Smith continued to work on the thermodynamics of the living organism, and Brillouin continued to call for physicists to study the exciting new "life principle." Although they did not develop a friendly reading of Schrödinger's "other laws" passage that persuaded them to engage in interdisciplinary work, neither were they turned away from interdisciplinary work by the text.

Conclusions

By coming to understand more about how Schrödinger's *What Is Life?* functioned rhetorically, we can offer a solution to the disputes that surround this book. Scholars who disagree about the value and meaning of the monograph tend to segregate into two main camps. Scientists such as Linus Pauling and historians such as Edward Yoxen focus on the cognitive content of the book and conclude that it fell short of scientific standards or that it worked on a limited community of now-forgotten scientists. Critics such as Evelyn Fox Keller focus on the social context in which the book appeared and conclude that it worked by way of social constructs such as political timing and social authority. The first group of scholars fails to explain the motivational impact of the book on a large number of successful scientists; the second group fails to explain what it was about the *content* of the book that allowed it to become an effective social force. As in my study of Dobzhansky's influence on the evolutionary synthesis, the rhetorical perspective connects diverse scholarly perspectives. Recognizing

52. Sydney Smith, "Rainbow Trout *(Selmo Irideus)*," 376.
53. Brillouin, "Life, Thermodynamics, and Cybernetics," 554–55, 565.

the importance of the "cognitive content" of the book, the rhetorical reading scrutinizes the arguments presented and asks serious questions about how those arguments may have made sense to the scientists who read them. Recognizing the importance of "social context," the rhetorical reading explores the political motivations that led those readers to accept one argument or another.

This solution will not be acceptable to everyone. Undoubtedly, some will still insist on describing the way in which Schrödinger's arguments did or did not fit into the progressive cognitive development of modern science, while other will insist that it was *kairos* and *ethos,* rather than particular argumentative strategies, that made Schrödinger's book so effective. And insofar as they argue that cognitive content and social context are important dimensions of criticism, they will be right. But a rhetorical perspective argues that each is only half right. A fuller understanding of Schrödinger's book would recognize the book's motivational content—namely, the way in which the text employed techniques such as polysemy to offer an inspirational appeal to individuals with different intellectual, disciplinary, and ideological commitments.

A rhetorical perspective also offers a solution to the other major scholarly dispute about this book by identifying polysemy as a rhetorical strategy employed (consciously or unconsciously) by Schrödinger to achieve his interdisciplinary purpose. Different scholars read the "other laws" passage in different, and indeed contradictory, ways. Rather than side with one scholar against another, a rhetorical analysis of the text and intertext shows that all three readings were supported by the text, all three interpretations were made by the people who read it at the time, and all three meanings contributed to the persuasive effect of the book.

III Edward O. Wilson's *Consilience: The Unity of Knowledge*

6

The Controversy over Sociobiology: Scholars Offer Conflicting Explanations

Dobzhansky's 1937 *Genetics and the Origin of Species* was designed to build a bridge between the disciplinary communities that studied genetics and natural history. Schrödinger's 1944 *What Is Life?* helped scientists span the somewhat larger gap between physics and biology. In 1998, Edward O. Wilson's *Consilience: The Unity of Knowledge* also attempted to inspire interdisciplinarity. But rather than seek to unite two areas of study within biology, or two areas of study within the natural sciences, Wilson envisioned a more ambitious connection across what C. P. Snow has called the "two cultures" divide. Wilson's book undertook "the attempted linkage of the sciences and humanities" (*WC*, 8).[1]

Wilson's Purpose

Early in his book, Wilson defined *consilience* as a type of interdisciplinarity, "a 'jumping together' of knowledge by the linkage of facts and fact-based theory across disciplines to create a common groundwork of explanation" (*WC*, 8). Announcing that the bridge of "consilience" already does a fair job of connecting disciplines in the natural sciences, he proclaimed that it was now time to extend that bridge by "integrating knowledge from the natural sciences with that of the social sciences and humanities" (*WC*, 13).

> The explanatory network now touches the edge of culture itself. It has reached the boundary that separates the natural sciences on one side from the humanities and humanistic social sciences on the other. . . . There is only one way to unite the great branches of learning and end the culture wars. It is to view the boundary

1. Although he identifies the division here as one between the sciences and the humanities, it becomes clear elsewhere that Wilson sees the social sciences as closely connected to the humanities, and his goal is to unite both with the domain of the natural sciences.

> between the scientific and literary cultures not as a territorial line but as a broad and mostly unexplored terrain awaiting cooperative entry from both sides. The misunderstandings arise from ignorance of the terrain, not from a fundamental difference in mentality. (*WC*, 125–26)

This unexplored terrain between the two domains of knowledge was the subject of Wilson's book, just as the unexplored terrain between genetics and natural history was the subject of Dobzhansky's book and the unexplored terrain between physics and biology was the subject of Schrödinger's. Wilson asked his readers to imagine a "map" that "can be thought a sprinkling of charted territory separated by blank expanses that are of unknown extent yet accessible to coherent interdisciplinary research." He described his book as an attempt at "gap analysis," in which "the efforts of scholars to explore" those empty spaces were traced and readers were urged to undertake the intellectual challenge of exploring those spaces themselves (*WC*, 267–68).

Just like other texts in the "interdisciplinary inspirational" genre, Wilson's *Consilience* reviewed the literature and made arguments for professional reorganization; it was not primarily designed to support a novel scientific truth claim. Early in the book Wilson admitted that his "belief in the possibility of consilience beyond science and across the great branches of learning" is "a metaphysical world view," not a scientific truth claim, and a minority belief at that, "shared by only a few scientists and philosophers" (*WC*, 9). His intent in this book was to offer the "strongest appeal" for the linkage of the natural sciences with the social sciences and humanities, an appeal based on "the prospect of intellectual adventure and, given even modest success, the value of understanding the human condition with a higher degree of certainty" (*WC*, 9).

Also like other texts in the interdisciplinary inspirational genre, Wilson's *Consilience* argued that a connection between the two cultures would be fruitful for both sides. "The human condition is the most important frontier of the natural sciences," proclaimed Wilson. "Conversely, the material world exposed by the natural sciences is the most important frontier of the social sciences and humanities" (*WC*, 267). After connecting the two cultures, readers could look forward to solving "the central problem of the social sciences and the humanities, and simultaneously one of the great remaining problems of the natural sciences" (*WC*, 126). In addition, Wilson promised that scientists would find opportunities for discovery waiting to be seized (*WC*, 155), economists and other social scientists would see their predictions become more accurate as they harvested an abundance of new ideas (193, 206), the scholarly analysis and criticism of the arts would become more powerful (211), semiotics would be enriched (136), structuralists would achieve potentially fruitful results (154), and the liberal arts would be revitalized, "if the natural sciences can be successfully united with the social sciences and humanities" (269).

Earlier Attempts at Interdisciplinary Inspiration by Wilson

This appeal for consilience was not the first attempt Wilson made to unite the two "domains" that had long been separated. Several of the reviews that responded to Wilson's *Consilience* recognized his task and his arguments in this 1998 book to be similar to those of his 1975 book *Sociobiology: The New Synthesis.*[2] In chapter 1 of *Sociobiology,* Wilson had argued that because the discipline of sociology is similar to taxonomy and ecology, involved in the description of outermost characteristics, descriptions drawn from "unaided intuition, without reference to evolutionary explanations in the true genetic sense," it must be reshaped in the same way that taxonomy and ecology "have been reshaped entirely during the past forty years by integration into . . . the 'Modern Synthesis' ": "It may not be too much to say that sociology and the other social sciences, as well as the humanities, are the last branches of biology waiting to be included in the Modern Synthesis. One of the functions of sociobiology, then, is to reformulate the foundations of the social sciences in a way that draws these subjects into the Modern Synthesis."[3]

On reading *Sociobiology,* some audience members immediately recognized that it was an attempt at interdisciplinary synthesis written for two audiences. According to Princeton biology professor John Tyler Bonner, Wilson was saying to biologists, "[D]o you not realize that this is a subject of prime significance in biology today?" while simultaneously saying to the social scientist, "This is what the biologist has to offer; surely this must be of consequence to you and to the future of your study of man."[4]

Wilson's own description of his intent in writing *Sociobiology* confirms that, as with *Consilience,* one of his main goals was to inspire interdisciplinarity. Not long after writing *Sociobiology,* he described the exigency he faced as follows:

> Natural scientists have by and large conceded social behavior to be biologically unstructured and hence the undisputed domain of the social sciences. For their part, most social scientists have granted that human nature has a biological foundation, but they have regarded it as of marginal interest to the resplendent variations in culture that hold their professional attention.[5]

Wilson wrote *Sociobiology* in response to this exigency; he saw the potential of sociobiology to be in "its logical position as the bridging discipline between the natural sciences on the one side and social sciences and humanities on the other."[6] Hopeful that his book would overcome the

2. See, e.g., Carman, review of *Consilience,* 169; Anderson, "Recycling Sociobiology," 101; Rose, review of *Consilience,* 36; Harvey, "Further Steps," 451.
3. Wilson, *Sociobiology: The New Synthesis,* 4.
4. Bonner, "New Synthesis," 129.
5. Wilson, Introduction to *Sociobiology and Human Nature,* 11.
6. Ibid.

barriers between fields and open the door to interdisciplinary work, Wilson looked forward to "a great deal more imaginative research by both evolutionary biologists and social scientists"; he optimistically predicted that his book would herald the beginning of an exciting new human science drawn from collaboration "between biologists and scholars in the social sciences and humanities."[7]

Wilson was especially hopeful that social scientists would be motivated by his call to action.

> I hoped to make a contribution to the social sciences and humanities by laying out, in immediately accessible form, the most relevant methods and principles of population biology, evolutionary theory, and sociobiology. I expected that many social scientists, already convinced of the necessity of a biological foundation for their subject, would be tempted to pick up the tools and try them out.[8]

According to Wilson, his "intention in writing the final chapter of *Sociobiology*" was to make the connection between human behavior and biology "compelling enough so that social scientists would add sociobiology to the corpus of social theory. Sociobiology and the broader evolutionary theory of which it is a part appeared to me and to others to be the most direct bridge from the natural sciences to the study of human nature and thence to the social sciences."[9]

Years later, when reflecting on his intent, Wilson repeated the same themes. He had believed "the time ripe for the melding of biology and the social sciences." The appeal to interdisciplinarity in the last chapter of *Sociobiology* "was meant to be a catalyst dropped among reagents already present and ready to combine."[10] He had hoped *Sociobiology* would make evolutionary biology "attractive to a younger generation of researchers in the social sciences, who might then connect their field to the natural sciences."[11]

The Effect of Wilson's Interdisciplinary Appeals

But Wilson is the first to admit that his hopes remain, for the most part, unrealized. In his autobiography, Wilson concluded that his expectation that young social scientists reading *Sociobiology* would find the connection of their field to the natural sciences attractive "was desperately naïve"; they clung strongly to their nonbiological presumptions.[12] In fact, Wilson was "surprised—even astonished" by the strong initial reaction to

7. Wilson, foreword to *Sociobiology Debate*, xiv; Wilson, foreword to *Sociobiology and Behavior*, xii.
8. Wilson, Introduction to *Sociobiology and Human Nature*, 1.
9. Wilson, foreword to *Sociobiology Debate*, xiii.
10. Wilson, *Naturalist*, 336.
11. Ibid., 334.
12. Ibid.

Sociobiology by both biologists and social scientists, who showed "stiff resistance" to his interdisciplinary appeal.[13] A book that was meant to work as a catalyst uniting biology and the social sciences "spun out of control," resulting in a "ferocity of response" from the very people he thought would be excited by the new prospects it promised.[14]

After *Sociobiology*, with the help of Charles Lumsden, Wilson wrote two other books that offered support for a bridge between the two cultures.[15] In these books he discussed "gene-culture co-evolution" through the concept of "epigenetic rules," an integrative biological theory connecting the natural and the social sciences (a theory that he later revisited in *Consilience*); but Wilson found himself once again disappointed because his work received a number of unfavorable reviews from influential social scientists and biologists. Wilson "was worried, and puzzled" that his attempts to inspire scholars to cross disciplinary boundaries were not "persuasive enough to attract scholars," for he was convinced that the connection between the two cultures "is the central problem of the social sciences, and moreover one of the great unexplored domains of science generally."[16]

Consilience is Wilson's most lengthy appeal yet. Whereas *Sociobiology* included two chapters of appeals for bridging the gap between the natural sciences and the human sciences, and some of Wilson's other books offered a theory to support the bridge-building efforts, *Consilience* is wholly devoted to the goal of inspiring readers to integrate the two cultures. But in the short time since it was published, Wilson's *Consilience* has fared no better than his earlier appeals. A majority of scholars in the natural sciences, social sciences, and humanities who have written book reviews of *Consilience* have responded negatively to Wilson's latest appeal for interdisciplinarity. The philosopher Richard Rorty represents the sentiment of most scholars in rejecting Wilson's call to interdisciplinary action:

> My overall reaction to *Consilience* is that although advances in biology may someday have greater relevance to the behavioral sciences, and conceivably even to the policy sciences and the humanities, than they do now, we should nevertheless not get on the bandwagon Wilson is trying to set in motion. We should not beat our breasts about our sadly disunited culture. We should not take measures to increase awareness of recent advances in evolutionary biology among the academics, nor to break down barriers between disciplines.[17]

13. Wilson, Introduction to *Sociobiology and Human Nature,* 1–2; Wilson, foreword to *Sociobiology Debate*, xiii.

14. Wilson, *Naturalist,* 336.

15. Lumsden and Wilson, *Genes, Mind, and Culture;* Lumsden and Wilson, *Promethean Fire.*

16. Wilson, *Naturalist,* 353.

17. Rorty, "Against Unity," 37.

Readers have called Wilson's interdisciplinary erasing of the boundaries between the natural sciences and the humanities and social sciences "indefensible," a sermon "shaky in substance" that is based on "a distressingly flat account of disciplines and their boundaries."[18] They complain that his notion of consilience rides on a "dubious assumption" about the proper relation between disciplines, is grounded on purported links between natural science and social science that are "fatally weak," and in the end will "not neatly solve any of the problems he notes, especially in the humanities and social sciences."[19] Scholars have suggested that Wilson's appeal did more to threaten intellectual harmony than promote unification, and they have predicted that the book will not be "winning many converts to consilience."[20] In assessing the book, both scientists and scholars of the humanities have concluded that the idea of consilience promoted by Wilson is "absurd" or flat-out "wrong," and they have remained "unconvinced" by his vision.[21] Even those who said they once admired Wilson were "disappointed" and left "with a heavy heart" after reading the book, concluding that Wilson's proposal for disciplinary consilience could not be accepted by serious thinkers.[22]

According to Wilson, the resistance to his most recent appeal for interdisciplinarity once again caught him by surprise. "Frankly I'm rather surprised that this idea—or shall we say prophecy or projection—has met so much resistance."[23] But resistance has been the most common response to Wilson's call for a new synthesis across the boundary that separates the natural sciences from the humanities and social sciences.

Of course, there have been some who accept the vision of Wilson's *Consilience*.[24] Likewise, Wilson's earlier appeals for interdisciplinarity can hardly be dismissed as total failures. Many more people work today in the borderlands between the social sciences and the natural sciences than before Wilson began his campaign to unite the fields, and though some of these people are careful to distance themselves from the term "socio-

18. Todorov, "Surrender to Nature," 33; Kevles, "New Enlightenment," 12; Burnett, "Dream of Reason," 145.

19. Pope, "Scientist's Search," 1031; Hirsch, "Pitfalls of Heritability," 33; Elshtain, review of *Consilience*, 59.

20. Midgley, "Well-Meaning Cannibal," 24; Spalding, review of *Consilience*, 442.

21. Dupré, "Unification Not Proved," 1395; Rose, review of *Consilience*, 37; Dyson, review of *Consilience*, 205.

22. D. Jamieson, "Cheerleading for Science," 91; Eldredge, "Cornets and Consilience," 84, 86. For statements by some of the other scholars who reject Wilson's appeal for consilience see Anderson, "Recycling Sociobiology," 104; Gould, "In Gratuitous Battle," 87; Jones, "In the Genetic Toyshop," 15; Steadman, review of *Consilience*, 325; Bernstein, "Wilson's Theory of Everything," 65; Berwick, "All Together Now," 12; and W. Berry, *Life Is a Miracle*, 95.

23. Wilson, "Divisive Ideas," B2.

24. Statements by some of those who accept Wilson's main argument in *Consilience* can be found in B. Berry, "On Consilience," 96–97; P. Gross, "Icarian Impulse," 48–49; Forrest, review of *Consilience*, 1796; Doyle, "Biology Takes Center Stage," 65.

biology," and others identify an intellectual heritage that is conspicuously "Wilson-free," some pay tribute to the inspirational effect of Wilson's *Sociobiology*.[25] Unlike the inspirational appeals of Dobzhansky and Schrödinger, however, which were almost universally accepted by their target audiences, Wilson's attempts to motivate interdisciplinarity have been loudly rejected by a large portion of his audience. The books of Dobzhansky and Schrödinger resulted in consensus; the books of Wilson have resulted in controversy.[26]

When considering this state of affairs, the obvious question that arises is why Wilson's appeal was not as successful at achieving consensus as the other appeals that have been examined in this book. Scholars who have commented on Wilson's appeal, from its origins in *Sociobiology* to its most recent presentation in *Consilience*, offer conflicting explanations for why so many readers have rejected it. Foes of Wilson say that his interdisciplinary appeal has been less than fully successful because it is, quite simply, wrong. In contrast, supporters of Wilson say that he has not been as successful as he hoped because he has been undermined by critics who are blinded by their political agendas.[27] After reviewing these two explanations and identifying some reasons for not accepting either as an exclusive answer to the question, I offer a third explanation that is further developed in the next chapter.

Explanation 1: Wilson Is Wrong; The Cultural Divide Should Not Be Bridged

One way of explaining the resistance to Wilson's interdisciplinary appeal is to recognize it as the appropriate response to a foolhardy attempt to bridge disciplines that cannot or should not be brought together. Critics of Wilson who offer this explanation believe that his central inspirational claim (that the domains should be united) is wrong and therefore deserving of the harsh criticism it has received. They suggest that the chasm

25. Segerstråle, *Defenders of the Truth*, 313–20.

26. Recall that though there was a great deal of recent scholarly dispute about whether Schrödinger's book should be valued as a contribution to the history of science, there was never a dispute about the correctness of his inspirational claim that biologists and physicists should work together. The historical evidence examined in chapters 2 and 4 of this book shows that both Schrödinger's text and Dobzhansky's text were tremendously successful at motivating their primary audiences. In contrast, a look at the existing evidence shows that most of those responding to Wilson's *Consilience* have countered his call to interdisciplinarity.

27. Of course, these explanations mirror the explanations each side give about what motivated Wilson to make the appeal in the first place. Those who are opposed to Wilson say that he was driven by a hidden political agenda; those who support Wilson say that he was a clear-eyed scientist animated by the truth. As the rhetorician Greg Myers points out, scholars responding to Wilson treat him as "either the product of scientific progress or the product of ideological reproduction"; Myers, *Writing Biology*, 224. Charging your opponent with political motives and maintaining your own position as the "truth" is a general rhetorical strategy employed in both scientific and public disputes.

separating the natural sciences from the humanities and the social sciences is simply too wide to span, and a rejection of Wilson's call to bridge those two domains is a fitting response offered by those wise enough to recognize this fundamental truth.

Many of the statements made by critics of Wilson's *Consilience* support this explanation. For example, Rorty argues that each academic discipline does its job quite well, so there simply is no "need to bridge gaps among the natural sciences, the social sciences, the humanities, and the arts."[28] Unpersuaded by the examples that Wilson offered to build a prototype bridge between the domains, Rorty sees no compelling reason for blurring the very useful boundaries between the humanities and the sciences.[29] The philosophers Mary Midgley and John Dupré offer similar responses, pointing out that the types of questions asked in the natural sciences and in the humanities are simply too different to be combined in the way that Wilson proposed.[30] The biologist Stephen Jay Gould agrees, arguing that the humanities and the sciences are distinct domains that cannot be collapsed into one another.[31] The biologist Steven Rose goes so far as to argue that even the natural sciences should not be combined with each other, much less the sciences with the humanities. Although there is one world, he says, we should respect the fact that there are many different ways of knowing about it.[32] Echoing these arguments, various other scholars from both sides of the "two cultures" chasm have concluded that Wilson's call to action in *Consilience* was misguided: they maintain that the basic differences between the natural sciences and the human sciences are appropriate and will always keep these separate systems of knowledge from being unified in the way that Wilson hopes.[33]

Years before *Consilience* was written, many scholars who responded to Wilson's earlier attempts to inspire interdisciplinarity drew the same conclusion. For example, in an early review of *Sociobiology*, the biologist John Maynard Smith argued that Wilson's "big claims for the relevance of biology to sociology" are "exaggerated or unjustifiable."[34] In examining the new sociobiology proposed by Wilson, the biologist George Wald was struck by its "massive irrelevance" to the present human condition, which, he said, is better described through traditional social, economic, and

28. Rorty, "Against Unity," 29–30.

29. Ibid., 33–34.

30. Midgley, "Well-Meaning Cannibal," 23; Dupré, "Unification Not Proved," 1395. Midgley made a similar argument in response to Wilson's earlier work; see Midgley, "Rival Fatalisms," 26.

31. Gould, "In Gratuitous Battle," 88. Gould makes a similar argument in response to Wilson's earlier work. See Gould, "Sociobiology and Human Nature," 289–90.

32. Rose, review of *Consilience*, 37.

33. For example, see Todorov, "Surrender to Nature," 33; Burnett "Dream of Reason," 146–47; Dale Jamieson, "Cheerleading for Science," 91; Eldredge, "Cornets and Consilience," 85.

34. J. Smith, "Survival through Suicide," 496.

political sciences.[35] The philosopher Marjorie Grene complained that sociobiology was blind to the fact that different questions are asked by different fields, and she predicted that Wilson's "new synthesis" would soon join other failed visions in an early and well-deserved death.[36] Scholars from fields as diverse as chemistry, biology, anthropology, and economics concluded that sociobiology was not an appropriate bridge between the natural and the social because it failed to recognize that the two domains are simply too different to be united in the way that Wilson proposed.[37] They were especially forceful in concluding that the social sciences would be better off on their own than they would be if unified with biology.[38] Each of these critics rejected Wilson's claim that the natural sciences can or should be united with the social sciences and humanities. To them it was obvious that Wilson's failure to persuade them was due to his decision to champion a claim that is false.

Explanation 2: Critics Are Unable to See the Truth Because of Political Bias

The flip side of the explanation offered by readers who reject Wilson's call is the one offered by Wilson and his supporters; they say that Wilson's appeal for interdisciplinarity has not been as successful as it should have been because vocal and influential critics who are blinded by their political agendas are unwilling to recognize the truth of Wilson's message. According to Wilson and his supporters, there are several ways in which readers have been motivated by their own interests rather than by the validity of Wilson's claim that unity of knowledge should be achieved: critics are tragically shut off from seeing the truth by their political affiliations, ideological orientations, professional interests, and disciplinary dogmas.

Complaining about the audience's response to his first interdisciplinary appeal, Wilson said that he did not receive criticism on evidential grounds about why the social sciences could not be united with the natural sciences, but "received instead a political enfilade" that "blindsided" him.[39] According to Wilson, the "politically correct formula" that dominated on most college campuses favored sociocultural explanations for human behavior because that was the only way that people could be assured that social behavior is "almost infinitely flexible" and "that different cultures must be accorded moral equivalency." Wilson's proposal that the social sciences draw on biological explanations called these political assumptions

35. Wald, "Human Condition," 277.

36. Grene, "Sociobiology and the Human Mind," 219–20, 224.

37. Alper, "Ethical and Social Implications," 201–2; Freeman, "Sociobiology," 209; Barnett, "Biological Determinism," 152; Lewontin, "Sociobiology," 30.

38. Washburn, "Animal Behavior," 54, 57, 62, 71; Boulding, "Sociobiology or Biosociology?" 273; Leeds, "Language of Sociobiology," 190.

39. Wilson, *Naturalist*, 338.

into question by suggesting that "some forms of social behavior are probably intractable" and that what is good and true has intrinsic validity.[40] According to Wilson, radically liberal political groups such as Science for the People attacked him because they were afraid that their political ideology would be threatened; they illogically saw his proposal to be "inimical to either revolutionary socialism or the goals of distributive justice."[41] Wilson believes that the "stiff resistance" he faced came because he had been tampering with the fundamental "mythology" of Marxists and other leftist scholars.[42] He says that critics were driven by their partisan goals to engage in "self-righteous vigilantism" against his proposal, leveling a "political attack" that deliberately misinterpreted his claims in order to support their own interests.[43]

This story of why so many people were unable to accept Wilson's call to action has been retold by several of Wilson's supporters. According to Garrett Hardin, Marxist critics intimidated Wilson with a single-minded intensity of attack that went beyond a search for truth; Wilson's failure to persuade some readers could be attributed to the unfortunate fact that "the loyalty of the 'true believer' . . . can overwhelm the most powerful intellect."[44] Other supporters have more recently inveighed against the "politically correct intellectuals . . . [who] brought their entire arsenal to bear on Wilson" and the "angry critics" who made "preposterous" political accusations out of fear that their ideologies would be threatened.[45]

According to Wilson's supporters, the critics' openly partisan attacks were underpinned by some more deep-seated ideological stances. For example, some supporters postulate that readers had "gut reactions" against the idea of biological explanations for human social behavior because that idea seems to rob mankind of its humanity, or because people tend to be resistant to theories that demystify human behavior.[46] One supporter concludes that the "root cause of the animosity" against Wilson is the "fear that if human phenomena were really explicable . . . then man would no longer be free."[47] Another says that readers who were unable to see the potential in Wilson's proposal have minds that "cannot tolerate so understandable, so logically mechanistic a world" as the one Wilson described; like losers unable to accept the fact that nice guys finish last, some intellectuals are resistant to the hard truths that result from a

40. Ibid., 334–35.

41. Wilson, "Attempt to Suppress," 277–78.

42. Wilson, Introduction to *Sociobiology and Human Nature,* 2.

43. Wilson, "For Sociobiology," 60–61; Wilson, "Academic Vigilantism," 183, 187–90.

44. Hardin, "Nice Guys Finish Last," 186–87.

45. Trefil, "Must All That Rises Converge?" G3; Pinker, "Theory of Everything," paras. 7–8.

46. Holton, "New Synthesis?" 81; van den Berghe, "Bridging the Paradigms," 52; Alexander, "Evolution, Human Behavior, and Determinism," 4.

47. Lawson-Tancred, "Start of an Intellectual Battle," 40.

biological explanation of social behavior.[48] Critics of *Consilience* are "biophobic," says one supporter; they are driven by a deep-seated dualism that blindly denies "that laws of nature coming from science offer any true or useful explanation of human behavior and society, or provide us any guidance."[49]

Wilson's supporters have also suggested that readers unable to accept his call for interdisciplinarity are blocked from seeing the truth by their professional interests. As Michael Ruse explains, this sort of thing happens all the time when a science that seems "closer to the truth or more adequate" than another science "moves in on the domain" of that other science; "the scientists being invaded feel threatened and resentful."[50] As another supporter puts it, Wilson has triggered counterattacks because "massive institutional and doctrinal vested interests in the university subculture and beyond depend for their livelihoods on his being wrong."[51] A sociobiologist might describe the negative response of so many scholars as "a casebook study in territorial behavior."[52] The historian of science Gerald Holton says, "It is not surprising to find assertions of territorial claims in the replies to Wilson by members of other disciplines."[53] The adverse response of a number of intellectuals is due to their concern "about what they perceive to be grand imperialistic designs on their area."[54]

According to some of Wilson's supporters, the territorial behavior that kept so many scholars from accepting Wilson's call to action is not based exclusively on a desire to maintain the current balance of power between disciplines; it has deeper cognitive roots.[55] Social scientists have a set of "self-imposed intellectual blinkers" that keep them from accepting the burden of rapprochement with the scientific fraternity.[56] One supporter has even gone so far as to compare nonbiologist critics in a university audience to small-town creationists who pronounce evolution "eevilution"; both have been trained to exclude biology and genetics from their considerations of human behavior and culture.[57] Wilson blames this cognitive blockage on "the strength and power of the antigenetic bias that has prevailed as virtual dogma" in the social scientific professions.[58] The "sociocultural view favored by most social theorists" is a "deeply rooted philosophy" in the social sciences, not open to hypothesis testing.[59]

48. Hardin, "Nice Guys Finish Last," 192, 188.
49. P. Gross, "Icarian Impulse," 42–43.
50. Ruse, *Sociobiology: Sense or Nonsense?* 165, 168.
51. Lawson-Tancred, "Start of an Intellectual Battle," 40.
52. Hardin, "Nice Guys Finish Last," 184.
53. Holton, "New Synthesis?" 85.
54. Ibid., 84.
55. Ibid., 86.
56. van den Berghe, "Bridging the Paradigms," 35.
57. Alexander, "Evolution, Human Behavior, and Determinism," 5.
58. Wilson, Introduction to *Sociobiology and Human Nature*, 2.
59. Wilson, *Naturalist*, 334.

Prelude to a Rhetorical Reading

Whereas critics of Wilson believe that his proposal to connect the natural sciences with the humanities and social sciences is a misguided effort that should be rejected because it is incorrect, supporters of Wilson believe that he has a true vision that is being blocked by individuals unwilling to see past their own biases. From both points of view, the rhetorical approach that Wilson took when presenting his call to interdisciplinarity is irrelevant. One side believes he failed with so many because he was wrong; the other side believes he failed with so many because his readers were unmovable. Either way, there is nothing that he could have done to persuade his target audience.

But when we read between the lines of each explanation, we find some subtle hints that Wilson might have persuaded those who did not already agree with him if he had taken a different rhetorical approach. If we closely examine the remarks of Wilson's critics, we find that they are not blinded by bias against the very idea of interdisciplinary collaboration; in fact, many of them seek it. Midgley, who soundly rejects Wilson's particular proposal for interdisciplinarity, strongly believes that "the various provinces of our intellectual world" should work together usefully, and she looks forward to the day when Wilson writes a book "which will make real reconciliation possible."[60] Another critic countered Wilson's proposal even while dreaming of an interdisciplinary behavioral science containing elements of both social science and biology.[61] Others who strongly oppose Wilson's *Consilience* agree with him that there is no reason to continue the "tedious culture wars," and they applaud any attempt at mutual comprehension across the divide.[62]

A close look at critics' arguments against Wilson's appeal show that they are not philosophically opposed to the idea of a biological explanation for human social behavior. They seem more than willing to accept the sociobiological thesis that humans have a biological nature "and that this nature has something to do with why we like certain kinds of art and why certain social structures would not suit us."[63] They "have no doubt" that Wilson is right when he says we have epigenetic rules hard-wired into our brain in the course of evolution and that there are more of these than we currently suspect.[64] They are persuaded by the argument that "evolutionary biological history—as determined by our genes—has something to do with the human condition" and that learning about the biological

60. Midgley, "Well-Meaning Cannibal," 24. See also Midgley, "Rival Fatalisms," 26.

61. Washburn, "Animal Behavior," 279. This dream is shared by one of Wilson's supporters; see Ruse, *Sociobiology: Sense or Nonsense?* 165–93.

62. Rorty, "Against Unity," 36; Todorov, "Surrender to Nature," 33; D. Jamieson, "Cheerleading for Science," 90.

63. Dupré, "Unification Not Proved," 1395.

64. Rorty, "Against Unity," 32.

constraints facing humanity would be important.[65] Even one of Wilson's most vocal critics, Richard Lewontin, accepted the core of Wilson's argument when he said, "It is undoubtedly true that human behavior like human anatomy is not impervious to natural selection and that some aspects of human social existence owe their historical manifestations to limitations and initial conditions placed upon them by our evolutionary history."[66] Most of the critics of Wilson's proposal are not biophobic dualists but level-headed materialists who have no doubt that mental processes are thoroughly material.[67] They are eager to accept the idea of collaboration between the natural sciences and the humanities and social sciences and are willing to admit that human social behavior is guided by biological constraints. They are just unwilling to accept Wilson's particular vision of interdisciplinarity.

A close look at the statements made by critics of Wilson opens the possibility of interpreting Wilson's failure to persuade as a rhetorical one; so too does a close look at statements made by Wilson's supporters. Though quick to blame the disappointing reception to Wilson's appeal on the narrow-mindedness of the audience, supporters have hinted that the message of *Consilience* could have been better designed to persuade.[68] They admit that Wilson's criticisms of the social sciences "may seem unkind" and that he is not very tolerant of viewing phenomena through any lens other than his own.[69] They acknowledge that Wilson "has no problems producing withering denouncements."[70] They concede that Wilson's book can be perceived as an affront to scholars in the social sciences and the humanities, like a scientist in the faculty lounge bellowing "Anything you can do, I can do better."[71] They confess that even those who share Wilson's worldview "will find much to provoke them" in some of Wilson's stronger claims.[72]

Wilson has heard all this before; supporters who read his earlier books also recommended that he brush up on his understanding of effective rhetorical strategies in order to persuade others. Hardin remarked that Wilson "made a tactical error" in *Sociobiology* by implying an invasive threat to the social sciences and humanities; Wilson was "asking for" criticism,

65. Eldredge, "Cornets and Consilience," 85; Alper, "Ethical and Social Implications," 202; Gould, "In Gratuitous Battle," 88; Gould, "Sociobiology and Human Nature," 283; Washburn, "Animal Behavior," 74.

66. Lewontin, "Sociobiology," 29.

67. Rorty, "Against Unity," 36; Lewontin, "Sociobiology," 30; Grene, "Sociobiology and the Human Mind," 213.

68. Of course, Wilson's critics have done more than hint at this. I discuss their awareness of Wilson's rhetorical miscalculations in the next chapter.

69. Carman, review of *Consilience*, 170, 171.

70. Harvey, "Further Steps," 452.

71. G. Fisher, review of *Consilience*, 1455.

72. Pinker, "Theory of Everything," para. 11.

rallying the troops *against* his own program.[73] Holton acknowledged that Wilson did "little to calm colleagues in neighboring territories" by offering a "direct attack" against their disciplines. Although Holton said the opposition to Wilson was not fully explainable by Wilson's impolitic statements, he acknowledged that it was "triggered" by them.[74] An early favorable review of Wilson's *Sociobiology* even made a joke about it; this supporter wondered sardonically if Wilson's appeal to social scientists and humanities scholars "is the best way of luring disciples of these historic schools to read the book ushering in the new synthesis":

> Any pool player appreciates the "unwisdom" of calling one's shots in advance; so what are we to do with the flat statement [by Wilson] that ethology and comparative psychology are both destined to be cannibalized by neurophysiology and sensory physiology from one end and sociobiology and behavioral ecology from the other? As a onetime neurophysiologist, your present reviewer may look forward to meeting his new friends, the sociobiologists, across this not entirely festive dinner table but he wonders a little about the relevant others.[75]

In short, critics and supporters of Wilson have opened a space for a rhetorical explanation of his failure to persuade. Though critics maintain that Wilson failed to persuade them because his proposal was wrong, they are willing to admit that the idea of interdisciplinarity between the natural sciences and the humanities and social sciences is a good one, and they seem eager to hear a proposal that negotiates a path between those areas. Though supporters maintain that Wilson's critics were an unmovable audience, they are willing to admit that Wilson's impolitic rhetoric may have contributed to his failure to get through to that audience.

According to the philosopher of science David Hull, critics of Wilson would like nothing more than to equate Wilson's ideas with pseudoscientific theories that were accepted by scientists for a short time even though they were false (such as phrenology), and supporters of Wilson would like nothing more than to equate Wilson's ideas with scientific theories that were psychologically and politically hard to accept but true (such as Darwin's theory of evolution); but the best approach to understanding the conflict over Wilson's theory is to consider it from both perspectives.[76] It is neither a simple "truth" that Wilson was wrong nor mere "politics" that

73. Hardin, "Nice Guys Finish Last," 184.

74. Holton, "New Synthesis?" 85, 79.

75. Morison, "Biology of Behavior," 87.

76. Hull, "Scientific Bandwagon," 136–38. Hull compares Wilson's proposal to both theories because he maintains that "the truth of new theories *as they are originally set out*" (his emphasis) tells us little about the ultimate acceptance of those theories, which may be altered over time by scientists who are more or less rhetorically savvy. I am making a somewhat different point when I say that the particular way in which theories are originally set out has something to do with whether they are accepted by their audience.

kept his theories from being accepted. In a way, we can say that it was both: Wilson was wrong to take the particular approach to unification that he did, not because interdisciplinarity is a flawed idea but because his particular approach to unification was designed in a way that angered scholars who otherwise would have been open to his call to action. The rhetorical design of his appeal and the manner in which it was presented kept it from being as persuasive as it could have been. Chapter 7 will explore the rhetorical strategies he employed in *Consilience*, contrasting them with the strategies used by Dobzhansky and Schrödinger, in the hope of providing a more complete rhetorical explanation of why Wilson was not as successful as he could have been with his interdisciplinary appeal.

7

A Text Rhetorically Designed to Fuel Interdisciplinary Hostilities

A rhetorical analysis of Wilson's *Consilience* can offer a new way of understanding why his appeal for interdisciplinarity has not been as successful as he hoped it would be. It suggests that both opponents and supporters are right to a certain degree. Wilson *was* wrong, but only insofar as his particular appeal for interdisciplinarity employed poor unifying strategies, and Wilson's audience *was* unmovable, but only because Wilson designed arguments that cemented them into a position against him.

In this chapter, I argue that several rhetorical choices made by Wilson negatively influenced the outcome of his appeal for interdisciplinarity. First, Wilson's failure to inspire those who were not already in his camp was largely due to the fact that he employed a rhetoric of conquest, rather than the rhetoric of negotiation used by Dobzhansky and Schrödinger. Many of Wilson's language choices hardened undecided or critical readers against his call for a bridge between disciplines. Second, Wilson lost many of his readers because of the particular theory he chose to promote in his interdisciplinary appeal. In contrast with Schrödinger's use of strategic ambiguity regarding the "laws of physics," Wilson was explicit in his commitment to reductionism, a stance that set many of his readers against him. Third, Wilson employed a form of polysemy that did more to damage his own ethos than to smooth tensions between disciplines. In contrast with Dobzhansky's use of a subtle Aesopian form to negotiate the competing professional interests of the two groups he would bring together, Wilson made rhetorical choices that allowed his audience to recognize the polysemy of his text; this ultimately led to a rejection of his appeals as two-faced. After examining each of these rhetorical explanations for Wilson's failure to persuade so many of his readers, this chapter concludes by imagining what Wilson could have done to achieve more success with his interdisciplinary appeal. It is my belief that Wilson could have achieved consensus around a new form of "consilience" that connected the natural sciences with the humanities and social sciences, but the rhetorical construction of that appeal would have had to follow a blueprint that was very different from the one he chose in this book.

A Rhetoric of Conquest, Not Negotiation

In her book on the history, theory, and practice of interdisciplinarity, Julie Thompson Klein points out that metaphors used to describe relations between disciplines are typically drawn from geopolitics.[1] Disciplines are often imagined to be "territories" with "boundaries" that can be "protected" or "crossed." So it is not surprising that the language of interdisciplinarity is capable of taking on the countering geopolitical images of hostile imperialistic conquest or friendly international alliance. Although there were places in *Consilience* where Wilson described interdisciplinarity as mutually beneficial "trade" across the "boundaries" that separate disciplinary domains, a close textual analysis of the book shows that the overwhelming majority of metaphors used by Wilson to describe interdisciplinarity established an image of one territory dominating another through an expansionist war.[2] This image made it difficult for many in Wilson's audience to accept his call to action.

Metaphors and Analogies in Wilson's *Consilience*

Geopolitical metaphors appeared extensively in Wilson's *Consilience.* In some places, readers were encouraged to imagine a "bridge" between the natural sciences and the humanities and social sciences, a bridge that would cross the "borderland" that divides the two "domains" (*WC,* 136, 178, 193, 267). But Wilson did not center most of his geopolitical metaphors around the image of an industrial-era construction project that unites territories with a mutual transit system; instead, his most frequent metaphors in *Consilience* evoked frontier-era images of exploration and conquest. In his effort to inspire his audience to engage in new lines of research, Wilson talked of disciplinary "frontiers" that cry out for exploration; he asked his audience to fill out the "map" of the material world, which has only a "sprinkling of charted territory separated by blank expanses"; he called for a "Magellanic voyage" that would boldly circumscribe the whole of reality, leading to further exploration (*WC,* 267–68, 298). Drawing on the heroic image of the pioneer or explorer, Wilson envisioned the search for consilience as a drive to the new frontier. "There lies the high adventure for later generations, often mourned as no longer available. There lies great opportunity" (*WC,* 270).

Evidence in the book indicates that Wilson was self-conscious about this rhetorical choice. For instance, he said that his "placement of the Enlightenment founders in mythic roles of an epic adventure was inspired by Joseph

1. Klein, *Interdisciplinarity,* 77.
2. For the metaphor of trade, see *WC,* 204, 321.

Campbell's *The Hero with a Thousand Faces* . . . and its application to popular culture by Christopher Vogler in *The Writer's Journey: Mythic Structures for Screenwriters and Storytellers*" (*WC*, 300). Although Wilson only commented on this strategy in a footnote to his chapter on the Enlightenment, he used these themes of epic adventure throughout the book, placing modern scientists in those same mythic roles. It is likely that he chose these metaphors because he believed that when audience members could imagine themselves as heroes on an exciting adventure, they would interpret his call to action favorably.[3]

In an interview with the German magazine *Der Spiegel*, Wilson suggested that he wrote *Consilience* in the way he did because "scientists need faith and encouragement so they can start out like discoverers believing in new, undiscovered continents."[4] It is likely that Wilson believed he could promote excitement about interdisciplinary activity by portraying those allegiant to the methods of the natural sciences as leaders on a "quest" for consilience, drawn by the appeal of "intellectual adventure" to plunge "repeatedly into new terrain" (*WC*, 7, 9, 12). His book urged scientists to "accept the adventure," to go "beyond the horizon" and find out what lies there (*WC*, 13). Scientists were portrayed as the type who follow "a guidebook to adventure plotted across rough terrain" (*WC*, 6). According to Wilson, over "the last several decades, the natural sciences have expanded to reach the borders of the social sciences and humanities"; here the "great adventure" truly starts as the scientist meets his "severest test," his "ultimate challenge" (*WC*, 66).

Insofar as these metaphors encourage audience members to envision a new interdisciplinary activity as heroic and exciting, they would seem to serve Wilson's inspirational purpose quite well. They appear to be similar to Dobzhansky's "inspirational calls" and Schrödinger's "appeal to fame." However, the fact that the hero in Wilson's narrative of exploration is the natural scientist, not the social scientist or scholar of the humanities, is not likely to be lost on his readers.

Throughout *Consilience*, Wilson described scientists as heroic adventurers who seek to expand the domain of science into territory previously claimed by the social sciences and the humanities. Heralds of adventure such as Francis Bacon were driven by the "thrill of discovery" and an implicit trust in the power of science to explain everything; they turned "with united force against the Nature of things, to storm and occupy here castles and strongholds, and extend the bounds of human empire" (*WC*, 21, 24). These scientists announced that "there is a new world waiting . . . let us begin the long and difficult march into its unmapped terrain" (*WC*,

3. Vogler's book portrays the terms of the hero's adventure as "a welcome tool kit" that has "stood the test of time" in creating stories that are "dramatic, entertaining, and psychologically true." See Vogler, *Writer's Journey*, 13.

4. Wilson, "Apes," 252.

28). Wilson also identified some more recent scientists who embody the crusader's spirit, such as complexity theorists, "a band of audacious" explorers who believe that the discovery of their "grail" is "on the near horizon," and brain scientists, who are "pioneers" living in a "heroic period," sending frequent "dispatches from the research front" (*WC*, 88, 99, 109).

> They are risk takers, who compete with rival theorists for big stakes and are willing to endure painful shake-outs. They bear comparison with explorers of the sixteenth century, who, having discovered a new coastline, worked rivers up to the fall line, drew crude maps, and commuted home to beg for more expeditionary funds. And governmental and private patrons of the brain scientists, like royal geographic commissions of past centuries, are generous. They know that history can be made by a single sighting of coastline, where inland lies virgin land and the future lineaments of empire. (*WC*, 100)

Taking a page from the imperialist rhetoric of politicians such as Theodore Roosevelt and Albert Beveridge, Wilson depicted scientists as rugged men committed to the strenuous life. According to Wilson, the most successful scientist, like a sixteenth-century explorer, is "confident enough to steer for blue water, abandoning sight of land for a while. He values risk for its own sake. He keeps in mind that the footnotes of forgotten treatises are strewn with the names of the gifted but timid" (*WC*, 58). Scientists moving beyond their own familiar territory are "journeymen prospectors" dreaming of the "mother lode"; they "travel to the frontier and make discoveries of their own, and as fast as possible, because life at the growing edge is expensive and chancy" (38, 56). They are adventurers who "travel around the system" until they "hit a soft spot," then "press on to new questions, new systems" (54). They often move forward on "blind faith," recognizing that "the more forbidding the task, the greater the prize for those who dare to undertake it" (209). Their occupation is "exploration of the universe in concrete steps, one at a time. Their greatest reward is occasionally to reach the summit of some improbable peak and from there, like Keats' Cortez at Darien, look in 'wild surmise' upon the vastness beyond" (121).

The persuasive difficulty Wilson could expect to face with this language of scientific conquest is clear; if scientists traveling beyond their territorial borders are explorers, pioneers, or conquistadors, then the indigenous peoples who currently occupy that land must be envisioned as backward and in need of civilizing, or lazy and undeserving of such rich territory. The language Wilson chose to describe scholars in the social sciences and humanities suggested just that. Early in his book, Wilson set up a hostile relation between the heroic scientist and the backward native on whose land he treads. According to Wilson, although a few professional philosophers indict, charge, and hiss at him for addressing a subject they consider their own, the real reason humanity is unable to solve its most vexing issues is that "our political leaders are trained exclusively in the social sciences and humanities. . . . The same is true of the public intellectuals, the columnists, the media interrogators, and think-tank gurus" (*WC*, 11, 13). Luckily though, things are changing: philosophy "is a shrinking domain," according to Wilson, and a new age is dawning in which we all have the common goal "of turning as much philosophy as possible into science" (*WC*, 12).

Social scientists are the first to be overrun by the natural scientist's march of the flag. Social scientists may be striving to achieve the predictive capacity of the natural sciences, but they are failing when you consider their track record "in comparison with the resources placed at their command" (*WC*, 181). In contrast to the medical sciences, which are progressing rapidly, the social sciences "are snarled by disunity and a failure of vision"; they are "split into independent cadres" that speak different languages and rule only within their "chosen domain of space"; they are "easily shackled by tribal loyalty" and are "still in thrall to the original grand masters" (*WC*, 182, 191). Their best insights only work "in the same elementary sense that preliterate creation myths explain the universe" (183). This is because the social sciences are at an earlier stage of development than the natural sciences; failing to probe far enough, they lack the connection to theory that true scientists have (188–89). But whereas social theorists might "wish to keep the borders of their dominions sealed and the study of culture unroiled by the dreams of biology," natural scientists are "fortunately not so bound" (209).

Scholars in the humanities fare no better before Wilson's imperialistic march; according to Wilson, they have proved themselves to be inferior and in need of governing. Unlike brain scientists, who have moved into "the dark nether regions" of the brain to discover the activity of the mind at work, philosophers who believe the mind to be their "proper domain" discover that they "can travel only so far" without the help of science, and "in the wrong direction" (*WC*, 102, 96). Likewise, theoretically inclined critics of texts "have tried many avenues into that subterranean realm" of the human mind, but "in the absence of a compass . . . they make too many wrong turns into blind ends" (216). Ever since the Enlightenment scientists "failed to colonize ethics," the study of ethics has advanced little, leaving the most distinguishing and vital qualities of the human species "a blank space on the scientific map" (33, 254). Theology, which claims the subject of human meaning for itself, is "encumbered" by "Iron Age" implements, and is thus "unable to assimilate the great sweep of the real world now open for examination," while Western philosophy, encumbered by "professional timidity," has "left modern culture bankrupt of meaning" (269). In contrast to the adventurous scientists they study, "some philosophers of science have thrown up their hands, declaring that the borderlands between the natural and social sciences are too complex to be mastered by contemporary imagination and may lie forever beyond reach"; they look to "monsters that dwelleth in the Great Maelstrom Sea, and they sigh, *No hope, no hope*" (208–9).

Wilson saved what was perhaps his most damning language for the "ultimate polar antithesis" of science—postmodernism, a subversion that "has seeped by now into the mainstream of the social sciences and humanities" (*WC*, 40, 42). Postmodernists are not friendly natives but "hostile forces" who "fret upon the field of play" and against which courageous scientists must defend "their posts" (43–44). They make up "a rebel crew milling beneath the black flag of anarchy" (40). They "favor snideness and rage over hope and other uplifting emotions" as they proclaim that there is "no lodestar . . . no scientifically constructible map of human nature from which the deep meaning of texts can be drawn" (214–15). But they are an ignorant crew; for example, Derrida is like "a faith healer unaware of the location of the pancreas" (41). They are also lazy. Their belief that the world is too complex to be understood through the reductionism promoted by science "is the white flag of the secular intellectual," a "lazy" way of surrendering to fate (297). Insofar as social scientists follow the natural scientists, "they will succeed"; if they make the mistake of following the postmodernists, they will concede "a premature surrender . . . lazily" devaluing intellect (190).

At one point, Wilson even seemed to recognize the hostility toward scholars in the humanities and social sciences that was embedded in his rhetorical choices. But unashamed, he boldly acknowledged that his call to action could be interpreted as an imperialistic drive: "Call the impulse Western if you wish, call it androcentric, and by all means dismiss it as colonialist if you feel you must. I think it instead basic to human nature" (*WC*, 100).

A look at the responses of audience members indicates that many of them did not find this impulse basic to their nature. Wilson's decision to use imperialist metaphors as part of an inspirational call to promote interdisciplinary "consilience" between natural scientists and scholars of the social sciences and humanities was not the most successful rhetorical move he could have made. Rather than negotiate an interdisciplinary treaty, Wilson's language choices sounded the call for increased hostilities, which made collaboration across territorial boundaries even more unlikely.

Reception of Wilson's Language Choices

As might be expected, social scientists and scholars of the humanities were the most vocal in their rejection of Wilson's language of imperialist expansion. The historian of science D. Graham Burnett proclaimed that the "tragedy" of the book was that Wilson seemed to want "not quite constructive dialogue, but rather *lebensraum* for a science rampant."[5] Burnett pointed out that although Wilson claimed he wanted to build a "bridge" between the two cultures, it "is a kind of bridge one might eye with suspicion, for the message comes through clearly: the humanities and the social sciences represent science's last frontier. Let us build a bridge, he effectively proposes, and take over your island."[6] Other scholars in the social sciences and humanities who criticized Wilson's book echoed the language of hostile invasion in their description of his project. One noted that Wilson's consilience sought to relate all the disciplines of knowledge "under the banner of natural science," with a "challenge" to the hard sciences to "extend" their principles "into the domain of the social sciences and the humanities."[7] Another complained that "Wilson's model of unity is not collaboration across disciplines but a hostile takeover of the humanities and social sciences by the natural sciences."[8] Yet another pointed out that Wilson's demand that "all others must accept the primacy of science, its interpretations, and perhaps even its language . . . does nothing to open the door to dialogue"; according to this critic, even though Wilson says he is calling for a "truce" between the two cultures, "he turns right around in the next sentence and fires a shot to end the short truce . . . it is not a diplomatic first gesture."[9]

What may be somewhat more surprising is that in addition to sparking anger from scholars of the humanities and social sciences, Wilson's rhetoric of conquest was unfavorably received by scientists. Many of the scientists who rejected Wilson's call to action echoed his language of imperialistic expansion when offering their negative assessments of his book. The physicist Freeman Dyson pointed out that Wilson wrote "with undisguised contempt for the many practitioners of the social sciences—psychology, anthropology, sociology, and economics—who try to understand human behavior without reducing it to biology." This did not appeal to Dyson, who said he ultimately rejected Wilson's call to action "because I value the diversity of culture more highly than the unity of science, the rebelliousness of people more highly than the consilience of ideas. To me, science is only one of many ways of exploring the human landscape, without any overriding authority over the others."[10] The biologist Steven Rose said that Wilson's "passion to colonize the universe for physics" embarrassed him. "No scholar, I believe, has the right to be so disrespectful of other fields of knowledge without at the least having the grace to try to appreciate how and why they approach their complex fields of enquiry. The contempt he shows for their work makes me, as a fellow biologist, blush for

5. Burnett, "Dream of Reason," 144.
6. Burnett, "View from the Bridge," 214.
7. Pope, "Scientist's Search," 1027.
8. D. Jamieson, "Cheerleading for Science," 90.
9. Spalding, review of *Consilience*, 442, 445. See also W. Berry, *Life Is a Miracle*, 30–38.
10. Dyson, review of *Consilience*, 205.

my subject."[11] Other scientists who rejected Wilson's appeal portrayed him as a "campus imperialist" laying "siege to the humanities in the name of biology" or leading a "raid on the humanities."[12]

Of course, there were some readers who were stimulated by Wilson's attack on the social sciences and humanities. The historian of science Charles Gillispie admitted that he shared "the author's predilection for both the Enlightenment and science and his distaste for postmodernism."[13] The philosopher Paul Kurtz agreed with Wilson that science has "expanded the frontiers of knowledge"; he found Wilson's book "bold and provocative" because Wilson, like Kurtz himself, "seeks to defend the sciences" against a "bevy of postmodernist deconstructionists, cultural relativists, ecofeminists, Afrocentrists, neo-Marxists, philosophical Feyerabendians and Kuhnians and Latourian social constructionists . . . [who come] from the faculties in the humanities, the arts, and the social sciences."[14] A medical doctor who reviewed Wilson's book proclaimed that "former science majors" will find a "delicious pleasure" in reading it since they will be "watching one of their own outshine" students and faculty in the humanities; they will find here "the exhilarating spectacle of postmodernism being slashed with biting wit and unconcealed contempt, no quarter asked or given."[15] The United Kingdom's Chief Scientific Officer, Robert May, admitted that in reading Wilson's book, he could not help but love "the clear, calm and often cruel phrases that drive [Wilson's] lance through black knight after black knight."[16]

But even though there were some who savored the mental image of a scientific attack on the social sciences and humanities, the vast majority of scholars responding to Wilson's language of conquest complained that it was offensive or embarrassing. Wilson should not have been surprised by this response. After all, his earlier attempt at bridge-building in *Sociobiology: The New Synthesis* was rejected by many for the same reason. In that book, he proposed a view in which "the humanities and social sciences shrink to specialized branches of biology."[17] He also imagined ethics being "removed temporarily from the hands of the philosophers and biologized" and psychology being "cannibalized" by neurobiology, which would establish a new "set of first principles for sociology."[18] The metaphors of imperialistic conquest were not as pervasive or consistent in his earlier work, but the hostility that Wilson felt toward the humanities and social sciences was clear to his readers. Responding to this text (and

11. Rose, review of *Consilience*, 36–37.
12. Lanier, "Biology Rules," 84; Eldredge, "Cornets and Consilience," 85.
13. Gillispie, "E. O. Wilson's *Consilience*," 282.
14. Kurtz, "Can the Sciences Be Unified?" 47.
15. G. Fisher, review of *Consilience*, 1455.
16. May, "Attempt to Link," 98.
17. Wilson, *Sociobiology: The New Synthesis*, 547.
18. Ibid., 562, 575.

others that Wilson wrote soon after), they complained that he took a "thoroughly denigrative attitude towards these [social] sciences and the humanities," displaying "an obsession with eliminating or emasculating" them and treating them as "the enemy" to be "attacked."[19] They noted that his purpose seemed to be to "subsume" or strip away the social sciences and humanities, and they viewed this as "bleak vision."[20] They focused on his image of death for the fields that failed to submit, and they considered it an appeal "not calculated to endear its author with all his readers."[21] As the rhetorician Greg Myers put it, "Wilson's disciplinary imperialism had as much to do with the reception of the book [*Sociobiology*] as did any of its claims or implications"; its "aggressive passages . . . would tend to alienate anyone coming to them with even slightly different views."[22]

As I pointed out in chapter 6, even the biggest supporters of Wilson recognized his rhetorical error in treating the social sciences and the humanities as territories to be conquered, or entities to be cannibalized, by the intellectually superior culture of science. If Wilson's supporters are right about detractors rejecting Wilson's thesis because of their selfish desire to defend their intellectual territory against what was perceived to be a hostile threat, then Wilson's own rhetorical choices did much to strengthen and confirm their negative perceptions. If, on the other hand, Wilson's critics are right that Wilson's thesis was rejected because it was so obviously wrong, perhaps what made it seem so unreasonable was the fact that Wilson claimed intellectual hegemony for one field, rather than suggesting that intellectual benefits could be gained for both sides from collaboration across disciplines. It is likely that Wilson could have developed a more reasonable thesis about the need for interdisciplinarity if he had dropped his image of the natural sciences conquering the territory of the social sciences and humanities and spent more time developing his less frequent, but more promising, metaphors of trade or collaborative exploration.

In my analysis of the rhetoric of Dobzhansky and Schrödinger, I showed that metaphors can be used to overcome intellectual barriers and effectively unite competing disciplines. Recall that Dobzhansky used a metaphor that compared genetic difference to geographical space. This served to alter the thought patterns of both audiences he would bring together; it encouraged geneticists who looked at gene frequencies to begin thinking of populations occupying real landscapes, and it encouraged naturalists who looked at populations in real landscapes to begin thinking of gene frequencies. Likewise, Schrödinger used metaphors to reciprocally influence the thought patterns of the two audiences he would bring

19. Leeds, "Language of Sociobiology," 164, 184, 195.

20. Alper, "Ethical and Social Implications," 197, for "subsume"; Schneewind, "Sociobiology, Social Policy, and Nirvana," 239, for "bleak vision."

21. Mackintosh, "Proffering of Underpinnings," 336. See also Beach, "Sociobiology," 116; M. Gregory, "Epilogue," 293.

22. Myers, *Writing Biology*, 214, 243.

together. His linguistic choices reversed the standard metaphors of the two disciplines he would unite, subtly suggesting to physicists that the biologist's living organism was accessible to mechanistic methods while also subtly suggesting to biologists that the physicist's atom was animated with life-like force. When we contrast these strategic uses of metaphor, each of which balanced and reversed the cognitive presuppositions of the two audiences they would unite, with Wilson's one-sided metaphor of conquest, it becomes even more clear why Wilson's book was less successful than theirs.[23]

So why did Wilson choose to so fully develop the aggressive metaphor of expansionist conquest in *Consilience?* He had plenty of evidence from the reception to *Sociobiology* that a hostile rhetoric would hurt his purported goal of building a bridge to "unite the two great branches of learning and end the culture wars"; and even if he somehow overlooked this evidence, surely he could have guessed that a rhetoric of conquest would make it hard for his readers to do as he hoped and "view the boundary between the scientific and literary cultures not as a territorial line but as a broad and mostly unexplored terrain awaiting cooperative entry from both sides" (*WC,* 126).

There is no way to know the mind of an individual rhetor or to conclusively determine the cause of his or her rhetorical decision making. But a strong argument can be made that Wilson's rhetorical choices in *Consilience* were influenced by an implicit theory of persuasion that was drawn from his sociobiological theory.

Explanation for Wilson's Linguistic Choices

The pervasive and sometimes self-conscious language of imperialist expansion in *Consilience* was, most likely, a deliberate choice. I believe that Wilson decided to use a rhetoric of conquest that was almost certain to sacrifice the potential assent of social scientists and scholars in the humanities because he thought that this approach would do the most to persuade natural scientists to engage in interdisciplinary activity.

The implicit theory of persuasion under which Wilson was operating when he applied the metaphor of conquest was drawn from his sociobiological theory. According to sociobiologists, most scholars who try to understand human behavior pay "too little attention to the properties of the real brain, which is a stone-age organ evolved over hundreds of millennia and only recently thrust into the alien environment of industrialized

23. This is not to imply that Dobzhansky and Schrödinger never used metaphors of war. In fact, as I pointed out in chapter 3 of this book, Dobzhansky invoked a metaphor of conflict when discussing the interaction of evolutionary forces. However, Dobzhansky and Schrödinger never used metaphors of conquest to describe the relationship *between the disciplines they would unite.* Instead, when discussing the competing interests of the communities they would bring together, they chose metaphors that would negotiate differences and encourage peaceful interaction.

society" (*WC*, 208). Wilson's sociobiological theory postulates that the human brain is influenced by "epigenetic rules" that bias learning and decision making, and these rules "were shaped during genetic evolution by the needs of Paleolithic people" (*WC*, 223). "Systematic logico-deductive thought," and the science that comes from it, are "product[s] of the modern age and [are] not underwritten by genetic algorithms" (208, 262). In contrast, the impulse to colonize is "basic to human nature" (100). According to Wilson, "territorial expansion and defense" is hereditary; it is an "epigenetic rule" that works as one of several "algorithms that function across a wide range of behavioral categories" (170–71, 172). The rationality of science may come hard to our Paleolithic brains, but the passion of intragroup support and intergroup hostility is basic to our nature. Given this understanding of human decision making, is it any surprise that Wilson, in an attempt to persuade his audience to devote themselves to interdisciplinary study, chose to use metaphors that invoked the deep-seated human desire for territorial expansion?

Other scholars have pointed out that Wilson has a tendency to think of the relation between disciplines in terms of a competitive Darwinian struggle for existence, marked by species survival or extinction.[24] In *Consilience*, Wilson displayed this tendency when he spoke of his own sociobiological concept as part of the Darwinian contest of ideas and when he predicted that his idea would eventually be the victorious survivor (*WC*, 44, 53). Recognizing that metaphors of conquest come easily to a naturalist trained in the linguistic tradition of Darwinian evolutionary theory, we can partially explain Wilson's rhetorical choices as residues of his disciplinary culture. I believe, however, that Wilson's persistent use of metaphors of conquest in *Consilience* was not just an artifact of his training as a naturalist. It was a principled decision to draw on the epigenetic rules rooted in our Paleolithic genes; Wilson believed that by appealing to our human nature, he could develop an inspirational call that would, like archetypes in mythic narratives, "live in our hearts" (*WC*, 28).

At one point in the book, Wilson imagined our Paleolithic ancestors as a people who loved their home ground but recognized that "beyond lies opportunity for expansion and riches" in lands guarded by "neighboring people—poisoners, cannibals, not fully human" (*WC*, 232). Wilson believes that the deep genetic history that made our Paleolithic ancestors want to defeat their barbarian enemies still guides our actions today; like them, we are most highly motivated by the desire for territorial expansion. In the context of this theory of persuasion, Wilson's decision to use the language of conquest in his inspirational appeal makes perfect sense.[25]

24. See, e.g., Myers, "Every Picture Tells a Story," 258; Midgley, "Rival Fatalisms," 24.

25. Of course, another way that the sociobiologist might put this would be to say that Wilson was guided by his own Paleolithic desire for conquest when he designed his appeals for interdisciplinarity.

More significant, the fact that so many scientists rejected Wilson's territorial appeal suggests that there is something wrong not only with his rhetorical approach but with the theory that grounds it. Wilson is fond of claiming that the question of whether his theory or a competing one is correct is an entirely "empirical" one (*WC*, 143, 179, 216, 264). The rejection of his appeal offers empirical evidence that this particular epigenetic rule does not guide human behavior. Wilson was mistaken in his assumption that these appeals would motivate his audience to accept his call to action.

At the end of this chapter, I discuss some alternative metaphors that Wilson could have used to inspire the sort of interdisciplinary activity he claimed to seek, metaphors that draw from a somewhat different theory of how human behavior is most effectively motivated. But before I do that, I wish to examine some of the other rhetorical choices Wilson made that kept his *Consilience* from achieving the effect that he was seeking.

An Explicit Commitment to Reductionism

Closely tied to Wilson's imperialistic rhetoric was his decision to explicitly promote the concept of reductionism. Recall that reductionism is the belief that all of science can ultimately be reduced to relatively simple deterministic physical laws.[26] Then consider the fact that the most frequently quoted passage in Wilson's *Consilience* is the following: "The central idea of the consilience world view is that all tangible phenomena, from the birth of stars to the workings of social institutions, are based on material processes that are ultimately reducible, however long and tortuous the sequences, to the laws of physics" (*WC*, 266).[27] Unlike Schrödinger's much-quoted passage about the "laws of physics," which was interpreted in contradictory ways by audience members who had different attitudes about reductionism, Wilson's passage was interpreted uniformly as an extreme reductionist statement, and it was rejected by almost all who quoted it.

Although reductionism drives some very successful scientific enterprises today, it is still a contested principle in the academy. Few in the mainstream believe in nonmaterial vitalist forces, but many believe that mechanistic explanations of complex phenomena that reduce those phenomena to simple deterministic laws are inadequate and should not be privileged over other types of explanation. In fact, a great number of those who responded to Wilson's *Sociobiology* (and to his other early arguments for interdisciplinarity across the "two cultures" divide) said that the thing

26. A brief definition and history of reductionism was provided in chapter 5 of this book.

27. Scholarly reviews that make direct reference to this passage include: Bernstein, "Wilson's Theory of Everything," 65; Berwick, "All Together Now," 12; Costanza, "One Giant Leap," 487; Hirsch, "Pitfalls of Heritability," 33; Jones, "In the Genetic Toyshop," 15; Midgley, "Well-Meaning Cannibal," 23; Todorov, "Surrender to Nature," 29–30; Werner, review of *Consilience*, 44.

that most hurt Wilson's argument was his commitment to reductionism. Even scholars who were generally supportive of Wilson's thesis suggested that he made a tactical and methodological error when claiming that ethical, social, and political phenomena must eventually be reduced to the molecular level.[28] Those who opposed Wilson's thesis were even more forceful in their critique of his overconfident belief in the power of reductionism.[29]

One would think that this response to his early work would have caused Wilson to back away from his claim that interdisciplinary activity requires the reduction of one discipline's explanations to those of another. But Wilson was actually more insistent about his promotion of reductionism in the pages of *Consilience* than he had been in his earlier texts. He used it to ground his claim that the territory of the humanities and social sciences should be given over to the natural sciences.

Reductionist Statements in *Consilience*

In the first few pages of *Consilience,* Wilson identified his commitment to reductionism by giving it a name. Invoking the Ionian reductionist and "founder of the physical sciences" Thales of Miletus, Wilson labeled his belief in reductionism "the Ionian Enchantment." Wilson said that an epiphany had led him to this "conviction, far deeper than a mere working proposition, that the world is orderly and can be explained by a small number of natural laws" (*WC,* 4).[30] Later in the book, Wilson defined reductionism as "the study of the world as an assemblage of physical parts that can be broken apart and analyzed separately" (*WC,* 29).[31] Reductionism involves the folding of "the laws and principles of each level of organization into those at more general, hence more fundamental levels" (55). "Total consilience" is the "strong form" of reductionism, "which holds that nature is organized by simple universal laws of physics to which all other laws and principles can eventually be reduced" (55).[32] Recognizing that some would read his book and charge him with "conflation, simplism, ontological reductionism, scientism" and other similar sins, Wilson pleaded "guilty, guilty, guilty" (11). He portrayed himself as wholly

28. Hardin, "Nice Guys Finish Last," 184–85; M. Gregory, "Epilogue," 292–93; Barkow, "Sociobiology," 181–82, 184–85.

29. Barnett, "Biological Determinism," 152; Freeman, "Sociobiology," 211–14; Gould, "Sociobiology and Human Nature," 289–90; Grene, "Sociobiology and the Human Mind," 216–21; Leeds, "Language of Sociobiology," 163–64; Lewontin, "Sociobiology," 30; McShea, "Gene-talk about Sociobiology," 187–88.

30. Wilson said he took this name for his commitment to reductionism from Gerald Holton.

31. See also *WC,* 54.

32. William Whewell's definition of consilience as a "jumping together" of knowledge across disciplines, which Wilson quotes when he first introduces the term, does not mention hierarchies or the reduction of one level of explanation to another more fundamental level (*WC,* 8). So the connection of consilience with reductionism is Wilson's own invention.

attached to the belief that phenomena could be described best by being reduced to the most basic laws of physics.

Throughout *Consilience*, Wilson argued that science itself is committed to reductionism. According to Wilson, science was the "engine" of the Enlightenment only because the scientifically inclined "agreed that the cosmos is an orderly material existence governed by exact laws" (*WC*, 22). Enlightenment scientists saw that the cosmos

> can be broken down into entities that can be measured and arranged in hierarchies, such as societies, which are made up of persons, whose brains consist of nerves, which in turn are composed of atoms. In principle at least, the atoms can be reassembled into nerves, the nerves into brains, and the persons into societies, with the whole understood as a system of mechanisms and forces. (*WC*, 22)

Wilson acknowledged that though reductionism "may seem today the obvious best way to have constructed knowledge of the physical world," it was hard for humanity to grasp at first; "Chinese scholars never achieved it" (*WC*, 30). But once achieved, it grew more and more powerful. Reductionism after the Enlightenment had an "unbroken string of successes during the next three centuries" (30). Today we know that "reductionism is the primary and essential activity of science" (54).

In Wilson's words, the "great success of the natural sciences has been achieved substantially by the reduction of each physical phenomenon to its constituent elements, followed by the use of the elements to reconstitute the holistic properties of the phenomenon" (*WC*, 134).[33] Biology has only recently become a "mature science" because biologists "have refined reductionism into a high art and begun to achieve partial syntheses at the level of the molecule and organelle" (90–91). Biologists now recognize that an "organism is a machine, and the laws of physics and chemistry . . . are enough to do the job, given sufficient time and research funding" (91). Likewise, the only root metaphor that would make psychology into a natural science is the idea of human beings as machines, an image that would allow psychologists to recognize that "mind" is bound by the laws of physics (42, 118). Similarly, according to Wilson, only reductionism will allow economics to become a mature science:

> The full understanding of utility will come from biology and psychology by reduction to the elements of human behavior followed by bottom-up synthesis, not from the social sciences by top-down inference and guesswork based on intuitive knowledge. It is in biology and psychology that economists and other social scientists will find the premises needed to fashion more predictive models, just as it was in physics and chemistry that researchers found premises that upgraded biology. (*WC*, 206)

33. See also *WC*, 211.

In the pages of Wilson's book, those who are not committed to the reductionist agenda are, almost by definition, those who are not scientists. Wilson ridiculed scholars in the social sciences and humanities because they consider reductionism "a vampire in the sacristy" and place an "anathema" against it (267, 211). The believers in the "Standard Social Science Model" who view culture as "an independent phenomenon irreducible to elements of biology and psychology" were described by Wilson as backward; they see the world from an "upside down" perspective (188). In addition to being characterized as superstitious or stupid, those opposed to reductionism were portrayed by Wilson as "lazy" or cowardly (297).

Wilson characterized science as reductionist and thus successful and the social sciences and humanities as antireductionist and thus unsuccessful; then he used this characterization to ground his call for an imperialist move by the sciences. He proposed that explanation by reduction can and should "be achieved across all levels of organization and hence all branches of learning" (*WC*, 71).

Despite Wilson's vivid characterizations, a look at the response of readers shows that his attempt to equate science with reductionism (and nonscience with antireductionism) was a misunderstanding of the culture of science and a miscalculation of what constitutes a powerful argument before a diverse audience. Scientists and nonscientists alike rejected Wilson's call for "total consilience" on the grounds that it was reductionist and therefore philosophically, pragmatically, and empirically inadequate.

Reception of Wilson's Reductionism

Scientists were almost unanimous in contesting Wilson's portrayal of reductionism as the essence of successful science. According to the computer scientist Robert Berwick, "reductionism has never succeeded," despite what Wilson thinks, and it is unlikely to succeed in the future; we should not climb into Wilson's "narrow Procrustean bed of reductionism."[34] The biologist Steven Rose, who had complained about Wilson's commitment to reductionism earlier, labeled it "old-fashioned," a belief not held by contemporary thinkers who recognize that although "we live in one world," there are "many, non-reducible, different but hopefully compatible ways of knowing about it." Quite simply, said Rose, "the Enlightenment claim was wrong"; we cannot understand our brains by physics alone, but must look at them as parts of bodies that have developmental and cultural histories.[35] Jerry Hirsch, a scientist whose own work on behavior-genetic analysis should have made him appreciate Wilson's call to action, concluded that the book's "purported links

34. Berwick, "All Together Now," 12.
35. Rose, review of *Consilience*, 36, 37.

between the laws of physics and the cultural behaviour of human beings are fatally weak."[36] Other scientists rejected Wilson's thesis because his epistemological commitment to reductionism was "incomprehensibly silly"; because Wilson made a logical error when he forgot that the whole can sometimes be more than the sum of its parts; and because, "for all practical purposes," the long and torturous reductions to the laws of physics that Wilson proposed are so long and torturous as to render them impossible to traverse.[37]

Scholars in the social sciences and the humanities agreed with their colleagues in the natural sciences that the reductionism of Wilson's *Consilience* was its biggest flaw. The sociologist Ullica Segerstråle, who once chided Rose for uncharitably critiquing the reductionism of Wilson's earlier work, had to admit that Wilson went too far with his "hyper-Enlightenment quest" in *Consilience*.[38] Richard Rorty used an analogy to demonstrate why Wilson was wrong in believing that a reductionist explanation is always best.

> Human beings, like computers, dogs, and works of art, can be described in lots of different ways, depending on what you want to do with them—take them apart for repairs, re-educate them, play with them, admire them, and so on for a long list of alternative purposes. None of these descriptions is closer to what human beings really are than any of the others. Descriptions are tools invented for particular purposes, not attempts to describe things as they are in themselves, apart from any such purposes. Our various slowly evolved descriptive and explanatory vocabularies are like the beaver's slowly evolved teeth and tail: they are admirable devices for improving the position of our species. But the vocabularies of physics and of politics no more need to be integrated with one another than the beaver's tail needs to be integrated with its teeth.[39]

Mary Midgley made a similar argument with the metaphor of a mapbook: the plurality of different forms of explanation is like "the plurality which we find at the front of our atlases where many maps of the world—political, physiographic, climatological, and the rest—confront us without implying that we live in many worlds."[40]

Readers pointed out that their rejection of Wilson's reductionism did not imply, as Wilson seemed to think, a commitment to vitalism, mysticism, or a belief in extraphysical forces; instead, it was simply a reasonable

36. Hirsch, "Pitfalls of Heritability," 33.

37. Eldredge, "Cornets and Consilience," 85; Bernstein, "Wilson's Theory of Everything," 65; Gould, "In Gratuitous Battle," 87.

38. Segerstråle, *Defenders of the Truth*, 290–91; 361–64.

39. Rorty, "Against Unity," 28–29.

40. Midgley, "Well-Meaning Cannibal," 23–24.

recognition that different forms of explanation can be valuable.[41] Critics argued that Wilson's desire to unify all knowledge through reductionism "rides on the dubious assumption that there is only one kind of truth," an assumption that "has increasingly been rejected" in the philosophy of science because "handwaving and bald assertion aside, there have been few successful attempts at carrying out the hypothesized reductions."[42]

Readers who have cited Wilson's *Consilience* in their own research or in their own calls for action have not taken up its reductionist quest. Researchers who cite Wilson's *Consilience* approvingly support a more restricted form of consilience, arguing for connections between different parts of a single discipline or for a "smaller-range" synthesis of the biological sciences themselves.[43] Others cite Wilson's *Consilience* only to reject its particular version of integration between fields.[44]

Even some of Wilson's strongest supporters became critics when commenting on his commitment to reductionism. Paul Kurtz insisted that despite Wilson's claims, "it is important that we continue to pursue holistic explanations for more complex systems of organizations, particularly in psychology and the social sciences where we need to deal with phenomena encountered in the contexts under analysis."[45] Robert May insisted that some of the behaviors that Wilson tried to explain by reducing them to sociobiology were better explained in economic or Freudian terms.[46] And Charles Gillispie pointed out that since even physics is finding reductionism inadequate, it is unlikely that Wilson's reductionism is the answer for social sciences such as economics.[47]

The fact that some of Wilson's supporters did not let his reductionism spoil their view of his larger point about interdisciplinarity indicates that for them, it was not a fatal flaw in the design of his argument. However, the fact that so many who rejected the book said they did so because they could not accept Wilson's reductionism indicates that his decision to focus on this commitment was a grievous mistake that turned many potential converts away from his message of the need for interdisciplinary connections.

If we contrast Wilson's explicit commitment to an extreme reductionist agenda with Schrödinger's strategic ambiguity regarding reductionism,

41. Ibid., 23; D. Jamieson, "Cheerleading for Science," 91. See also Gould, "In Gratuitous Battle," 87.

42. Pope, "Scientist's Search," 1031; D. Jamieson, "Cheerleading for Science," 91. See also Todorov, "Surrender to Nature," 30.

43. Forehand, "Clinical Child," 178; "Consilience, Complexity, and Communication," 983.

44. Hales, "Problem of Intuition," 146; Bunk, "Is Science Religious?" 11; Ellis, "Nancey Murphy's Work," 603.

45. Kurtz, "Can the Sciences Be Unified?" 49.

46. May, "Attempt to Link," 98.

47. Gillispie, "E. O. Wilson's *Consilience*," 283.

we understand one reason why Wilson was less successful than Schrödinger. Because Schrödinger designed a book that could be read in more than one way, he was able to inspire those who believed in the power of reductionism as well as those who believed that scientists should seek to discover "new laws" irreducible to the current laws of physics. In contrast, Wilson adopted such an extreme and unambiguous reductionist position that even those who were nominally committed to the idea of reductionism had to admit that Wilson's vision took the matter too far.[48]

Equivocation Rather Than Productive Polysemy

After examining Wilson's bold imperialist metaphors and his explicit commitment to reductionism, one might begin to think that Wilson was an insensitive extremist, more interested in inciting opposition than achieving persuasion. There are, however, some signs of another voice straining to be heard in Wilson's book, a voice that is less hostile and potentially more attractive to undecided or skeptical audience members.

For example, despite an abundance of imperialist metaphors comparing scientists to sixteenth century explorers, Wilson assured "scholars in the humanities" that "scientists are not conquistadors out to melt the Inca gold. Science is free and the arts are free . . . the two domains, despite the similarities in their creative spirit, have radically different goals and methods" (*WC*, 211).[49] While researchers in the two domains continue to pursue their different goals, they should seek exchange since "science needs the intuition and metaphorical powers of the arts," just as "the arts need the fresh blood of science" (*WC*, 211). This idea of collaborative exchange connected well with the trade, bridge, and interzone investigation metaphors that appeared in a few places in the book.[50]

48. Another useful point of contrast can be seen when we compare the strategies employed by Wilson with those employed by Dobzhansky. Dobzhansky explicitly argued that the two areas he would combine represent different levels of explanation that should be respected; he did this in order to persuade the two sides that their paradigms (Mendelianism and Darwinism) were not conflicting but were in fact compatible with each other. He also employed arguments designed to increase the mechanistic thinking of naturalists, *and to decrease the mechanistic thinking of geneticists*, in an effort to connect the conceptual patterns of his two audiences. Rather than develop a one-sided argument for the reduction of all explanations to a single type, he helped his readers see the need for a *synthesis* of these different levels of explanation. In comparison, Wilson's one-sided reductionism is less well designed to encourage synthesis and more likely to increase conflict between areas.

49. Though this passage is somewhat accommodating, the careful reader will note the switch in terms here, from "humanities" to "the arts." Scholars of the humanities are unlikely to feel reassured until Wilson recognizes that the *humanities* and the sciences are separate domains. This Wilson is unwilling to admit; it is his belief that the arts are autonomous, but the interpretation and theory of the arts (typically the domain of the humanities) should come from "stepwise and consilient contributions from the brain sciences, psychology, and evolutionary biology" (*WC*, 216).

50. I have mentioned these elsewhere. See the first few pages of this chapter and of the previous chapter.

In addition to providing images of interdisciplinarity that were less antagonistic than his imperialist metaphors, Wilson included passages where he acknowledged that his philosophical commitment to extreme reductionism might turn out to be an intellectual mistake. After describing total consilience as the "strong form" of reductionism, Wilson admitted that this worldview "could be wrong. At the least, it is surely an oversimplification":

> At each level of organization, especially at the living cell and above, phenomena exist that require new laws and principles, which still cannot be predicted from those at more general levels. Perhaps some of them will remain forever beyond our grasp. Perhaps prediction of the most complex systems from more general levels is impossible. (*WC*, 55)

In a couple of other passages, Wilson repeated this concession that despite his hopes, his reductionist vision might be wrong, and those who reject it might be right (*WC*, 209, 268).

These disarming statements, and a few others like them, represented an approach that appeared to be more accommodating to the beliefs and interests of a diverse audience, and were thus more well adapted to the interdisciplinary inspirational goal Wilson had set for himself. However, in competition with the more extremist voice of Wilson as imperialist and reductionist, these infrequent passages were drowned out. For those who recognized the more moderate claims at all, they served only to mark Wilson as a crafty equivocator. Most of those who commented on these more moderate statements expressed frustration with Wilson's two-faced rhetoric. According to Tzvetan Todorov, there were actually two versions of consilience that appeared in the book, the "hard" version and the "soft" version. The former set up a biological determinism that reduced human characteristics, meaning, and emotions to the laws of physics and attempted to assimilate the humanities and social sciences into their scientific superiors.[51] The latter revealed "the more agreeable Wilson."

> For this other Wilson, physics does not adequately explain life, nor does biology explain culture. Genes prescribe no particular cultural trait, and conclusions drawn from heredity are always extremely risky. . . . Culture is much more responsible for our differences than genes: "The culture of the Kalahari hunter-gatherers is very distinct from that of Parisians, but the differences between them are primarily a result of divergence in history and environment, and are not genetic in origin." This is why human affairs should not be left to physicists or to biologists: scientists remain habitually "poorly informed about the rest of the world," and "many accomplished scientists are narrow, foolish people."[52]

51. Todorov, "Surrender to Nature," 29.
52. Ibid., 33.

Facing these two contradictory readings of the same text, Todorov concluded that Wilson was not in fact a divided man; "there is one Wilson who writes, a little cunningly, on two levels. His hard version is the sensational one, designed for the newspapers. His soft version is the prudent one, which enables him to respond to objections by retorting 'But that is exactly what I'm saying!' "[53]

Several other critics recognized the way in which the term *consilience* seemed to shift in meaning from extreme reductionism to a mild form of connectivity between different kinds of explanations.[54] Said the historian of science D. Graham Burnett, "this oscillation accounts for the irenical tone of some passages in the book," juxtaposed with "irruptions of what might be called high church oozism (it all goes back to the primordial ooze!)."[55] Noting a similar oscillation when Wilson set out "a serious indictment" then followed it with a "quite modest claim," Jeremy Bernstein complained that "again and again in this book," he felt that Wilson was yanking his chain.[56]

Critics have long complained about Wilson's penchant for equivocation. A book review of *On Human Nature* (1978) remarked that the text demonstrated a "tension between the rash and the cautious, the outrageous and the dull, the provocatively reactionary and the orthodoxly liberal," leading to "an all-pervading confusion as to the nature of the arguments being advanced and the conclusions that may legitimately be drawn from them."[57] Examining Wilson's *Sociobiology*, the rhetorician Greg Myers pointed out that Wilson "tends to intersperse long stretches of cautious suggestion and qualification with a few brash overstatements."[58]

Ironically, in his more mature work, Wilson tends to intersperse long stretches of brash overstatement with a few cautious suggestions and qualifications.[59] But the result is the same: audience members recognize the polysemy in his text and are frustrated by it.

Wilson's *Consilience* is like other texts in the interdisciplinary inspirational genre in that it resulted in a form of polysemy. Unlike the more

53. Ibid.

54. Burnett, "Dream of Reason," 144; Dupré, "Unification Not Proved," 1395; D. Jamieson, "Cheerleading for Science," 90.

55. Burnett, "Dream of Reason," 144.

56. Bernstein, "Wilson's Theory of Everything," 64. Bernstein actually used the metaphor "pulling my string," which he introduced with a story about a toy. The sentiment is the same; it roughly translates into a feeling of frustration caused when someone who thinks he is clever plays with you at your expense.

57. Mackintosh, "Proffering of Underpinnings," 338.

58. Myers, *Writing Biology*, 220.

59. Contrast the increasing hostility to the social sciences and humanities in Wilson's later works with Dobzhansky's increasing friendliness toward natural history in the later editions of *Genetics and the Origin of Species*. Ironically, Wilson becomes less friendly to the less powerful field as time goes on, whereas Dobzhansky became more friendly toward the less powerful field in his synthesis.

successful texts, however, *Consilience* was constructed in a way that allowed readers to recognize the multiple meanings it contained, and once recognized, this polysemy worked against the interests of the author. The rhetorical difficulty with Wilson's text is made clear when we compare his oscillation between a "soft" and a "hard" consilience with Dobzhansky's Aesopian form, which included a conspicuous division and subtle unification of the fields he would unite. Recall that Dobzhansky distinguished between the morphological method of the naturalist and the physiological method of the geneticist, then immediately said that his book was going to be concerned with the latter. After making this brief statement, however, he proceeded to incorporate the naturalist's methods into his discussion of "evolutionary dynamics," which took up *more than half of the book.* This subtle, but substantive, turn toward the professional interests of the less prestigious disciplinary group allowed members of that group to ignore the earlier statement that seemed to reject their methods.

Wilson, like Dobzhansky, spoke to two groups, and used two voices. But unlike Dobzhansky, Wilson made repeated antagonistic statements rejecting the methods of scholars in the humanities and the social sciences, statements that could not easily be forgotten when "soft" Wilson made his rare appearances. Also, unlike Dobzhansky, Wilson did not follow his rejection of the methods of the humanities and the social sciences with a sustained and substantive incorporation of those very methods into the perspective he promoted in his book. Without this substantive counterbalance to language rejecting the methods of these fields, his more moderate statements seemed to be empty words; they collapsed under the force of his more fiery attacks. Because readers recognized the shift between "hard" Wilson and "soft" Wilson, and because they had no reason to dismiss his more hostile words in light of a more subtle and extensive move that honored the methods of the community he had wronged, they concluded that Wilson was equivocating, and they decided to reject his more moderate statements as clumsy political moves that did not represent his true beliefs.

What Wilson's *Consilience* Could Have Been

It is not difficult to imagine a *Consilience* that would have avoided the problems that Wilson encountered. Rather than focus on a rhetoric of conquest, it would have developed metaphors that evoke the sort of negotiation needed for successful interdisciplinary collaboration. Rather than dismiss the methods of the humanities and the social sciences by promoting an explicit and extreme reductionism, it would have employed a strategic ambiguity regarding the ultimate need to reduce everything to the "law of physics." Then, with the voice of "hard" Wilson muted, the voice of "soft" Wilson could have done the strategic work necessary to bring scientists and scholars of the humanities and the social sciences together in collaborative activity across their disciplinary borders.

Alternative Metaphors

In one part of *Consilience*, Wilson complained that whenever journalists and college teachers describe a particular disagreement in the academic community, they characterize it as "the clash of antipodean views," and "when the matter is drawn this way, scholars spring to their archaic defensive postures. Confusion continues to reign, and angry emotions flare." Instead of drawing the matter in this way, he said, we should "call a truce and forge an alliance" (*WC*, 188). Had Wilson taken his own advice, he might have described the natural sciences and the human sciences as groups that are collaborating and can benefit from further collaboration, rather than portraying them as cultures at war. One way he could have done this while still developing the excitement of the frontier metaphor would have been to speak about scientists and scholars of the humanities and social sciences as partners in the exploration of uninhabited interzone territories. There are at least two places in the book where Wilson drew just this sort of image (*WC*, 126, 267). Like his conquest metaphors, this metaphor of interzone exploration was designed to create excitement for the new study by setting it in terms of a heroic adventure. Unlike his conquest metaphors, however, this one did not imagine the subjugation of indigenous tribes; instead, it implied that equal partners were uniting in the exploration of uninhabited territory that lies between their respective nations. To make this metaphor work, Wilson should have used it more frequently, and he should have eliminated the pervasive images of conquest that established such a "clash of antipodean views" in his book.

Another alternative metaphor that Wilson could have used to suggest productive collaboration between disciplines is drawn from his own field of biology. When discussing the evolution of disciplines, rather than describe a bloody struggle for existence that leaves only one species surviving, he could have used the metaphor of commensalism or of mutualism. In nature, organisms do not always compete with each other "red in tooth and nail"; sometimes animals live in close relationships, with one species benefiting while the other experiences neither benefit or harm (commensalism), or with both species benefiting (mutualism). A metaphor that imagined the relation between disciplines as an instance of commensalism or of mutualism would have made it easier for audience members to develop positive feelings about Wilson's call for interdisciplinarity.

A similar biological metaphor would imagine the relation between disciplines as the connection between different parts of a single organism. At one point in *Consilience*, Wilson drew an extended analogy that envisioned humanity as the Greek hero Theseus and the world as the labyrinth into which the hero ventured. Ariadne's thread, which allowed Theseus to go to the center of the labyrinth to slay the evil Minotaur and then find his way back out again, was, in Wilson's analogy, the thread of consilience. The "deep interior" of the labyrinth represented the "nebula of pathways through the social sciences, humanities, art, and religion," while the gallery at the entrance to the labyrinth, the home base to which Theseus hoped to return, was the empirical knowledge of physics (*WC*, 67). This metaphor of conflict, with its assumption that the hero must slay the demon residing in the social sciences and humanities and then return to a safe home in the hard sciences, made it impossible for readers to imagine an interdisciplinarity that was anything other than a reductionistic attack from one intellectual culture on another. Imagine that instead of portraying the disciplines as different parts of the labyrinth, Wilson had compared the disciplines to different parts of a single organism. For example, like the parts of a tree, the disciplines are all connected in an essential way. In fact, one part cannot live on its own without the others; if any large part of the tree is chopped off, the rest of the tree is likely to die. Also, like the parts of a tree, the disciplines are arranged hierarchically, with the roots of physics connected to the trunk of biology, which is connected to the branches of the social sciences, and the leaves of the humanities. Though arranged in a hierarchy, one part is not really any more important than another; roots, trunk, branches, and leaves are all important, and all must work together for the functioning of the whole. This metaphor, like the labyrinth analogy, creates the image of essential connectivity and hierarchy between disciplines, but unlike the labyrinth analogy, it does not imply that one discipline is the site of home and victory, while others harbor something alien, evil, and destined to be destroyed.

A final alternative metaphor is that of trade. As with the interzone exploration metaphor, there are at least two places where Wilson's chanced upon this alternative himself (*WC*, 204, 321). Imagine what would have happened if Wilson had dropped the pervasive metaphors of conquest and replaced them with those of trade. Interdisciplinary collaboration could have been perceived as mutually beneficial exchange across disciplinary boundaries. Wilson could have even justified the use of this metaphor to himself by recognizing it as an appeal to a deep-seated, Paleolithic impulse. After all, Wilson believes that another one of the fundamental epigenetic rules that guides human behavior is the tendency toward "contractual agreement" (*WC*, 171).[60] Rather than base his theory of persuasion on the sociobiological belief that people are most effectively motivated by the appeal to territoriality, Wilson could have based it on the sociobiological belief that people are most effectively motivated by their "genetic propensity to form long-term contracts" (*WC*, 297). Had Wilson used the trade metaphor and encouraged his readers to launch their

60. Wilson points out that this hereditary ability to establish a mutually agreeable contract is a way for humans living in communities to ensure that selfish needs are met when interacting with others.

epigenetic programming for contractual agreement, rather than their epigenetic programming for territorial defense and attack, he may very well have had more success at motivating interdisciplinary work. Each side would have been assured that their own interests were being met in the collaborative venture, and neither side would have had to worry about becoming either colonized victim or imperialistic subjugator.

An Alternative to Reductionism

Metaphors of cooperative interzone exploration, or commensalism, or trade would have been practical tools for inspiring interdisciplinarity in Wilson's *Consilience* because they would have implied that scientists and their counterparts in the humanities and the social sciences can work together in an atmosphere of mutual respect. In order to use these metaphors, however, Wilson would have had to shift his perspective somewhat. Rather than dismiss scholars in the humanities and social sciences as ignorant, misguided, lazy, or primitive, and rather than belittle those who believe that the natural sciences are "one respectable intellectual subculture in the company of many" (*WC*, 186), Wilson would have had to demonstrate an appreciation for the intellectual work that is done by scholars who reside on the other side of the university campus.

One way for Wilson to demonstrate such an appreciation would have been to soften his commitment to strict reductionism. As long as Wilson was determined to fully reduce the humanities and social sciences to scientific explanation, he had to distort their purposes and belittle their accomplishments. Many of those who renounced Wilson's *Consilience* recognized this and complained that Wilson's commitment to strict reductionism showed that he did not truly understand the arts, the humanities, or the social sciences.[61]

Imagine what would have happened if, instead of disparaging these disciplines, he had showed respect for their intellectual enterprises and agreed that they should continue their own work even as they collaborated with scientists to develop consilient explanations. Several of his readers were eager to imagine just such an alternative. Although they were adamant about rejecting his idea that the totality of the humanities and the social sciences should be assimilated by the natural sciences, they were more than willing to accept a more "modest," but not insignificant, thesis that science can develop a sort of "anthropology of morals and aesthetics."[62] They could imagine a university in which "the various provinces of our intellectual world" worked together harmoniously, usually "exploring different regions of reality," but when sharing territory, always "respecting each other's differences as separate enterprises" and often collaborat-

61. Todorov, "Surrender to Nature," 31; Rorty, "Against Unity," 36; Kevles, "New Enlightenment," 12; Midgley, "Well-Meaning Cannibal," 23; Dupré, "Unification Not Proved," 1395; Berwick, "All Together Now," 12.

62. Gould, "In Gratuitous Battle," 87; Dupré, "Unification Not Proved," 1395.

ing to "coduce" explanations that draw on more than one discipline at the same time.[63] Some readers even imagined ways in which the unique methods of the humanities and social sciences could play a significant intellectual role in the development of consilient explanations. They called for a consilience that would run in both directions, with science helping to explain the arts and social systems, *and* the humanities and social sciences helping to explain our understanding of physical reality.[64] They said that science should not merely try to reduce other fields to its own explanatory structure but instead should expand its own landscape of thought, and they looked to a future in which "all forms of human inquiry (science included) will come to look very different indeed."[65] In short, while rejecting Wilson's strict reductionism, readers were open to imagining other forms of connection that would allow for interdisciplinary work between the natural sciences, the humanities, and the social sciences.

Before he wrote *Consilience,* Wilson had plenty of evidence that his audience was ready to be approached about interdisciplinary connections between the separate domains of knowledge, as long as those connections were not described in strict reductionist terms. After writing *Sociobiology,* he received much criticism of his reductionist strategy from readers who demanded a less severe approach. Several suggested that a better approach would be to forego claiming intellectual hegemony for natural science and instead make a less radical claim that it can contribute to the explanation of human social behavior.[66] Others suggested that a more effective appeal for interdisciplinary exchange would recognize that the social sciences and the humanities have something important to contribute to the collaboration as well.[67]

Perhaps Wilson chose not to use a more modest rhetoric because he thought that bold language was necessary to get the attention of his readers. The rhetorician John Lyne and the biologist Henry Howe once fantasized about an appeal from Wilson in which "the hedges had been more prominently displayed and the striking imagery of genetic determinism had been omitted," but they concluded that such an appeal would not have been considered noteworthy by its audience.[68] Others have suggested

63. Midgley, "Well-Meaning Cannibal," 24; D. Jamieson, "Cheerleading for Science," 90; Kurtz, "Can the Sciences Be Unified?," 49.

64. Rose, review of *Consilience,* 37; Anderson, "Recycling Sociobiology," 102

65. Berwick, "All Together Now," 12; Anderson, "Recycling Sociobiology," 104; Burnett, "Dream of Reason," 145

66. Alexander, "Evolution, Human Behavior, and Determinism," 3; Lewontin, "Sociobiology," 29–30; Simon, "Biology," 294; Depew and Weber, "Innovation and Tradition," 239.

67. Griffin, "Humanistic Aspects of Ethology," 251; Peter and Petryszak, "Sociobiology versus Biosociology," 75–76; Washburn, "Animal Behavior," 279.

68. Lyne and Howe, "Rhetoric of Expertise," 145.

that a text that gave "soft" Wilson more of a voice would have been intellectually acceptable but trivial or banal, and thus easily ignored or forgotten.[69] After comparing Wilson's strategy to the strategies taken by other, more successful appeals for interdisciplinary collaboration, however, I must disagree with these assessments. Dobzhansky and Schrödinger both constructed rhetorically sensitive calls for interdisciplinarity using "soft" appeals that could be read positively by people with different interests. Their texts were extremely popular, judged noteworthy and long remembered even though they did not set out a bold reductionism or use controversial attacks to draw attention to themselves. In fact, their success can be attributed at least partially to their strategic avoidance of controversy. By employing rhetorical techniques that negotiated the interests of competing disciplinary groups, they convinced both communities that they would benefit from collaboration. Had Wilson used a few devices from the rhetorical toolkits of Dobzhansky and Schrödinger, and had he put away strategies that served only to increase division and hostility, he would have been more likely to succeed at his goal of motivating interdisciplinary activity.

69. Rose, "'It's Only Human Nature,'" 169; Todorov, "Surrender to Nature," 33.

IV Speaking to Multiple Audiences

8

The Genre

With the case studies complete, we can now make comparisons that will allow us to draw some conclusions about the genre of scientific writing that seeks to motivate interdisciplinary activity. By identifying similarities and differences between the rhetorical strategies of these three texts, we can come to recognize which characteristics are necessary to the social action of the genre and which are only accidental attributes of a particular text. Since two of the case studies focused on highly successful books in this genre, and the third examined a somewhat less successful book, we can also draw some conclusions about which strategies of persuasion are most effective when seeking to inspire scientists to cross disciplinary boundaries.

The most essential characteristic of the interdisciplinary inspirational monograph is its purpose. Texts that were meant to forge alliances across fields should be recognized as fitting this genre, whether they were ultimately successful at achieving this goal or not.[1] Texts that were meant to speak to narrow disciplinary concerns, or to establish the truth of a new scientific discovery, or to secure a grant for a particular research project, or to serve one of the many other functions of scientific texts, are not a part of this genre.

To make more fine distinctions about the specific rhetorical form of texts in this genre, I will first compare and contrast the successful interdisciplinary texts by Dobzhansky and Schrödinger, making particular

1. For a text to fit in a genre, it need not be a success at achieving its goal. For example, the speech in which President Clinton admitted to an "inappropriate" relationship with Monica Lewinsky and attempted to explain his earlier failure to admit to that relationship is commonly referred to as an *apologia*, the genre of political speech in which one defends or justifies one's actions. Many of those who call this speech an *apologia* do not believe it successfully defended or justified the president's actions. Its failure to achieve its purpose does not exclude it from the genre.

note of the things that distinguish this type of text from the prototypical scientific text, the research report. Then I will discuss the ways in which Wilson's text both fits and does not fit the requirements for success in the genre.

Comparison of Dobzhansky and Schrödinger

A comparison of Dobzhansky's *Genetics and the Origin of Species* and Schrödinger's *What Is Life?* indicates that some specific features of the successful "interdisciplinary inspirational monograph" are the following: texts in this genre rarely report original research, concentrating instead on summary and synthesis of information that is already published; they are usually written by individuals who have some status in one or both of the communities to be joined; and they include appeals that are targeted to at least two different audiences. Let's examine each of these characteristics in turn.

Synthesis Rather Than Original Research

The first identifying characteristic is what most clearly distinguishes this genre from the prototypical scientific research report. Rather than develop and defend novel truth claims, both Dobzhansky's book and Schrödinger's book derived their theoretical and factual content from previous accounts. Dobzhansky relied heavily on the published work of Sewall Wright, and Schrödinger based his arguments on Max Delbrück's section of the "Dreimännerwerk." The purpose of the interdisciplinary inspirational monograph is to show how collaboration is a promising professional action, not to argue for the truth of an original scientific claim. So it is not surprising that both texts were designed to "popularize" existing evidence that supported the idea of crossing disciplinary boundaries. That evidence was relatively unknown to most members of the isolated disciplinary communities, but it was not *new* evidence, and since it was already published, it was not rendered through the conventional structure of a research report, which is designed to establish a scientific truth claim.

Though it is clear that both texts were synthetic works that were more concerned with inspiring action than with gaining intersubjective support for an original truth claim, it is equally clear that Dobzhansky was careful to popularize only the most factual, up-to-date information, and Schrödinger was not. Four years after Dobzhansky wrote the first edition of his book, he found it necessary to produce a second edition that included "substantial additions" and "a complete rewriting of some parts" in order to better "direct the attention of investigators toward facts which may otherwise be ignored or not noticed at all."[2] Ten years after that, he revised the text again, adapting it to a new post-synthesis situation in

2. Dobzhansky, *Genetics and the Origin of Species*, 2d ed., xiii.

which "inconsistent" conclusions and the lack of a "common language" were no longer bars to interdisciplinary collaboration.[3] In contrast, Schrödinger was not particularly careful to include only the most up-to-date information, and he never made substantial revisions to his book. He gave lectures about the subject in 1948 and again in 1952, but both times, he made exactly the same arguments that he made in the 1944 book.[4] He never took into account the fact that some of the evidence he used was no longer accepted by the scientific community, nor did he point to new evidence that further supported the move toward collaborative action.

This dissimilarity suggests that though the genre calls for the synthesis of existing theories and facts, there is much room for variance in the quality of the information presented. The success of Schrödinger's book proves that it is not necessary to offer only the most recent and the most reliable evidence to support the claim that interdisciplinary action will be fruitful. Since the central claim of the book is a political one, rather than a scientific one, it does not demand the same strict standards of evidentiary recency that a prototypical scientific text would require.

Of course, the success of Dobzhansky's book shows that fidelity to strict evidentiary standards doesn't hurt, either. Three reviewers praised his use of "up-to-date" information and offered that as an argument for why others should read his book.[5] One of the consequences of the fact that Dobzhansky showed concern for accuracy and Schrödinger did not was that certain "experts" in the new interdisciplinary space responded to their books differently. Wright was the scientist whose work Dobzhansky had popularized, and his praise for the book was high: "There is no other book which combines such an extensive review and analysis of the present state of knowledge on the genetics of variability within wild populations."[6] In contrast, when Max Delbrück, the scientist whose work Schrödinger popularized, commented on *What Is Life?* he developed a long critique that pointed out flaws in the author's description of the physical nature of the gene.[7] The contrast between the responses of Wright and Delbrück

3. Dobzhansky, *Genetics and the Origin of Species,* 3d ed., vii.

4. Yoxen, "Schrödinger's 'What is Life?'" 36.

5. Grüneberg, review of *Genetics and the Origin of Species,* 69; Mayr, review of *Genetics and the Origin of Species,* 300; Just, review of *Genetics and the Origin of Species,* 1105. Whereas Schrödinger's book didn't even have a bibliography, Dobzhansky's was twenty-nine pages long, and was mentioned by most of his reviewers. When Dobzhansky came out with a second edition, Sewall Wright approvingly pointed out that the bibliography had increased to forty-six pages! Wright, review of *Genetics and the Origin of Species,* 284.

6. Wright, review of *Genetics and the Origin of Species,* 283–84.

7. Delbrück, "What Is Life?" 370–72. Although Delbrück critiqued the book's flaws, he also predicted that it "will have an inspiring influence by acting as a focus of attention for both physicists and biologists." Evidently, he recognized that factual flaws would not keep the book from achieving its purpose.

suggests that when experts who were already doing boundary work read these books, they recognized the care with which the evidence was treated, and were likely to describe the book on those grounds.

But the fact that Schrödinger's book was just as successful as Dobzhansky's when it came to motivating interdisciplinary action reminds us that those experts were not the primary audience. The scientists who were *not* already working in the interdisciplinary zone were the target audience, and they were not disturbed by evidentiary mistakes because they did not yet have the expertise in the interdisciplinary area to recognize those mistakes. In short, an "interdisciplinary inspirational" monograph is primarily designed to synthesize existing information; it doesn't hurt for the text to be constructed with the most recent scientific knowledge, but it need not meet the same requirements for accuracy as a text that is addressed to a core group of experts who are adjudicating the truth of a scientific claim.

Distinguished Ethos

The second characteristic that distinguishes the interdisciplinary inspirational monograph from the prototypical scientific text is the requirement that the author be an established leader in one or more of the fields being discussed. Though a scientific truth claim need only be written by someone who has been properly socialized in the practices of the relevant scientific field, a claim that interdisciplinary action will yield productive results requires a sort of social authority that goes beyond that minimal requirement. To be believable, the author must show that he or she has an ability to recognize successful professional moves when he or she sees them. Both Schrödinger and Dobzhansky carried this level of ethos when they wrote their books. The former was something of a scientific superstar, a Nobel Prize winner whose eponymous equation was taught in every basic physics class. The latter was not as famous, but his already vast and growing publication record, and his connection to both the naturalist and the geneticist traditions, made him a figure with the required status.

The fact that both books were derived from invited lecture series shows that both authors had been regarded by their peers as having special standing in the scientific community. It also explains the "teacher" persona that both authors adopted in their books. In chapter 3 I described the simplifying techniques that Dobzhansky used to teach his audience about the complex mathematical theories of the population geneticists. A close look at Schrödinger's text shows that it used similar techniques to explain difficult ideas in biology and physics. Like Dobzhansky, Schrödinger was careful to define his terms, he plugged numbers into the equations he used, he described "thought experiments" to help the audience understand the more difficult concepts, and he employed metaphor and analogy to help his readers visualize ideas.[8]

There is one significant way in which the personae of the two teachers diverged, however. Though Dobzhansky wrote in the restrained voice of a scientist speaking to other scientists, Schrödinger wrote in the informal voice of a scientific popularizer. For example, throughout his book, Dobzhansky used a citation system to credit every scientific fact that he included. He also wrote in the passive voice and referred to himself in the third person when citing his own work. In contrast, Schrödinger rarely cited sources; he mostly used his footnotes for other purposes: to apologize for oversimplification, to retract a claim, to expand a claim, to define a term, and to describe a visual aid that was used in the lecture from which his book was derived. There were only three places where he used footnotes to actually cite sources.[9] In addition, Schrödinger adopted the active voice in much of his book, and he used personal pronouns throughout. He spoke about his own life, explaining how the book arose from a course of public lectures that he delivered before a packed audience and reminiscing about how he first approached the subject as a "naive physicist" (*SW*, 1, 4). In addition, Schrödinger imaginatively used himself as an example several times in the book: he talked about how cell division makes "a body cell of mine . . . only the 50th or 60th 'descendant' of the egg that was I"; he traced out his own family tree to show how one of his chromosomes stands an equal chance of having come from his grandfather, Josef Schrödinger, or his grandfather's wife, Marie, née Bogner; and he used himself as an example to explain the possibility that a child of his would exhibit a recessive mutation that he hypothetically carries (*SW*, 22, 26, 39). These personal comments created an informal persona, in contrast to the professional, distant, and precise persona that Dobzhansky exhibited in his book.

If one were to compare the books solely by the look of the page and the "sound" of the speaker's voice, it would be clear that they belong to different genres. Despite these differences, however, both books were designed to do the same thing and both worked the same effect on their audiences. The reason for the difference in voice probably had less to do with the constraints of the genre than with personal style and the different status each author carried in the scientific community. Although both had the leadership ethos that is a minimal requirement for the successful interdisciplinary inspirational text, their level of authority was different.

8. For an example of how Schrödinger is careful to define terms, see *SW*, 37–38; for examples of how he plugs numbers into equations, see: 16–17, 51–52, 62–63, 64–65; for an example of a thought experiment, see: 11–15; for a few of his most striking metaphors and analogies, see: 3, 21, 22, 41, 43, 61, 74, 76, 79.

9. Apologies: *SW*, 22, 35–36, 49; retractions: 2, 5, 24, 25, 56, 69; expansions: 9, 14, 21, 44, 61; definitions: 20, 21, 51; description of visual aid: 54; citations: 3, 44, 51.

Schrödinger was unquestionably an elder statesman of the scientific community; in contrast, Dobzhansky was a promising *new* leader in the genetics of natural populations. Consequently, Schrödinger could get away with an informal style and not damage his ethos as a Nobel Prize–winning physicist who had good advice to offer, whereas Dobzhansky had to work to maintain his ethos as a scientist with the proper standing to make claims about the future of genetics and natural history.

Dual Address

The final characteristic of the genre is that the text includes strategies designed to appeal to at least two disciplinary audiences. Schrödinger's book clearly addressed two audiences; he began with the explicit assurance that he spoke "to both the physicist and the biologist" (*SW*, 1). Dobzhansky's dual address was more circumspect. Though he never explicitly said who his audiences were, he appeared to speak directly to the geneticist and somewhat more indirectly to the naturalist.

As before, the difference between Schrödinger's style and Dobzhansky's style might have had something to do with the different situations they faced. Schrödinger was working to connect researchers from two major disciplinary structures. Although there was a clear power differential between the two disciplines, they were not directly competing entities; at the time, there was no serious fear that biology was going to removed from the universities if physics continued its recent increase in social power. Subsequently, Schrödinger could explicitly address both audiences in his call for new interdisciplinary research. In contrast, Dobzhansky was working to merge two more directly competing disciplinary communities. Genetics and natural history were both subsets of biology, and at the time Dobzhansky wrote, the former seemed to be well on its way to displacing the latter. Since the battle was more immediate and more intense, Dobzhansky had to be more careful with his negotiation strategies.

Despite differences in the explicitness of their dual address, when we examine the textual strategies used by Schrödinger and Dobzhansky, it becomes clear that both books were well designed to build bridges between conflicting theories and practices. For example, both texts applied strategies to assure their audiences that common ground existed between two disciplines. Schrödinger did this most effectively with apologies and retractions that assured biologists and physicists that they shared the scientific norm of precision. Dobzhansky created common ground by explaining how recent findings demonstrated compatibility between Mendelian genetics and Darwinian evolution. An even more interesting way in which these texts built bridges between different audiences is through the rhetorical strategies that I have called conceptual chiasmus and polysemy. The degree to which these two exemplars of the genre share this form of persuasion is remarkable.[10]

Dobzhansky used conceptual chiasmus when he introduced and extended Wright's map metaphor. By describing changes in gene frequencies as if they were movements on a landscape, he promoted a conceptual shift that created space for collaboration between two disciplinary communities. Geneticists who were used to thinking of gene frequencies under laboratory conditions were encouraged to think about them in terms of populations moving about in real space; naturalists who were used to thinking about populations inhabiting ecological niches were encouraged to think about the gene frequencies that are carried by those populations. In short, each group was directed to think about its own subject matter in a way that corresponded to the other group's worldview, and the novel combination of perspectives worked to promote collaborative study.

Schrödinger used this strategy when he crisscrossed common disciplinary linguistic practices. By animating the atom, he captured the attention of some biologists who would have otherwise found the study of physics to be a dull mechanical process; by mechanizing the organism, he assured some physicists that biological matter was accessible to their systems of analysis. Each group was made to believe that what they were most interested in studying would be found in greater degrees across disciplinary lines. As with Dobzhansky's use of metaphor, a crisscrossing of intellectual space directed the two audiences to think about their respective subjects in a way that promoted collaborative study.

The rhetorical strategy of polysemy also promotes collaboration by appealing to diverse audiences; it works by encouraging different interpretive communities to develop contradictory readings of a text. Dobzhansky developed this strategy when he adopted an Aesopian form in his opening chapter, with a blatant appeal to the geneticist and a more subtle appeal to the naturalist. Geneticists read the opening claim that the book "is not concerned with the morphological aspect of the problem" and concluded that Dobzhansky's purpose was to explain why geneticists should apply their superior talents to the study of evolution in natural populations, a subject matter that was being mishandled by naturalists. On the other hand, naturalists read that chapter's division between statics and dynamics as the promotion of a more equal merger of the two fields. In this

10. I should note that I did not come to these texts with the expectation that I would find these strategies at work. Although I have written an article about polysemy, I did so only after I discovered this rhetorical device in the texts by Dobzhansky and Schrödinger (at which time, I decided there was a need to further describe the theoretical basis of this strategy). Although it is likely that I was more sensitive to finding these strategies in one text after finding them in another, I believe the evidence I have offered from both text and intertext shows that these concepts were not force-fitted onto the case studies, but instead, existed in the artifacts themselves.

example of polysemy, two groups with differing levels of power in the university were persuaded by their conflicting interpretations of the text to take actions that would result in the crossing of disciplinary boundaries. Geneticists were made to believe that the time was right for a territorial advance; naturalists were made to believe that collaboration would result in the preservation of their disciplinary practices. Had the text been written in a way that did not allow these two different reading frames, it might have lost its interdisciplinary force: without the explicit rejection of the naturalist's methods, geneticists might have dismissed the call to action as the flank protecting maneuver of someone who was a loyalist to the traditions of natural history; without the implicit support of the naturalist's methods, naturalists might have felt the need to dismiss Dobzhansky's ideas as the improper trespass of another hostile geneticist.

An even more striking example of polysemy can be found in Schrödinger's book, when he says that "other laws of physics" may soon be discovered in biology. This passage was read in conflicting ways by at least five different interpretive communities. Some physicists were inspired by the idea that exciting new laws of physics would be found in the living cell, some biologists were reassured by the idea that biology would retain a unique character if new quasi-physical properties were required to explain living matter, some physicists were motivated by the idea that new reductionistic "order-from-order" laws would be found in the living organism, some physicists were reassured by the idea that biology could be thoroughly reduced to quantum mechanics, and some biologists were encouraged by the idea that the current laws of physics would suffice to explain the secret of life. Because the text could be read in radically different ways, each conflicting group believed that its own ideological and professional interests were being met by the text, and therefore, each could agree to the ultimate call for action toward which all the interpretations led. Without the strategic ambiguity that allowed these very different reading frames to surface, the appeal of Schrödinger's text would have been more restricted.

Polysemy is particularly useful in interdisciplinary inspirational discourse because it convinces each side that collaborative action is in *its own* best interest. Since the goal of this genre is to motivate action, not reach agreement about the truth of a knowledge claim, consensus about meaning is not a requirement. In political terms, the reasons for taking action do not have to be shared by two disciplinary communities for both to agree that collaborative action is desirable; in fact, the most politic text is often the one that speaks to the conflicting interests of disputants with a single appeal that is heard in two or more different ways.

Wilson's Participation in the Genre

Edward O. Wilson's *Consilience* had some of the same characteristics as the interdisciplinary inspirational monographs produced by Dobzhansky and Schrödinger. Like those books, his was designed to synthesize existing information rather than introduce a new scientific theory. Wilson even made this point explicit when he described his book as "an account of the efforts of scholars to explore" the blank spaces of interdisciplinary research (*WC*, 267–68). Wilson was well aware that the main point of his book was not a scientific claim, and therefore "cannot be proved with logic from first principles or grounded in any definitive set of empirical tests" (*WC*, 9). Instead, his claim that we should seek consilience across the great branches of learning was a call to action, an appeal to readers to cross disciplinary borders.

With respect to the accuracy of the information presented, Wilson's text is more like Schrödinger's than Dobzhansky's. Some individuals who were already working in the interdisciplinary space were dissatisfied with the information Wilson chose to present. For example, Jerry Hirsch, who described his own research as behavior-genetic analysis, complained that Wilson "commits a fundamental error" in his discussion of genetics and human diversity and proves his understanding of the matter to be "inadequate to the point of being grossly misleading."[11] Walter Truitt Anderson, a political scientist who described himself as a convert to the biopolitical movement, complained that Wilson ignored several important works in this area, as well as in the area of bioeconomics and in the psychology of morals.[12] Others noted avoidable errors both in Wilson's scientific examples and in evidence drawn from the humanities and social sciences.[13]

The tremendous success of Schrödinger's text, however, despite its own series of avoidable errors suggests that lack of accuracy was not the reason Wilson's text was unable to fully achieve its goal of inspiring interdisciplinarity. The effectiveness of Schrödinger's book suggests that this genre requires a synthesis of existing information but does not require flawless precision in its review of the most recent knowledge. Audience members who are not looking for factual errors in this sort of book are not likely to find them or be deterred by them. For example, supporters of Wilson who reviewed the book did not even recognize the errors; they believed that Wilson was rigorous and current in his use of information from many different disciplines.[14] And most of those who criticized Wilson on the grounds of accuracy of information made a point of rejecting his main thesis for other reasons.[15]

11. Hirsch, "Pitfalls of Heritability," 33.

12. Anderson, "Recycling Sociobiology," 102, 104.

13. See, e.g., Berwick, "All Together Now," 12; Kevles, "New Enlightenment," 12; Todorov, "Surrender to Nature," 30, 31; Pope, "Scientist's Search," 1030, 1032; Jones, "In the Genetic Toyshop," 15.

14. See, e.g., Pinker, "Theory of Everything"; Harvey, "Further Steps Toward One Culture," 451.

15. Berwick, "All Together Now," 12; Kevles, "New Enlightenment," 12; Pope, "Scientist's Search," 1031–32; Jones, "In the Genetic Toyshop," 16.

With respect to the second characteristic of the genre, that the author be a distinguished leader in one or more of the fields being united, Wilson clearly fit the bill. Even reviewers who rejected Wilson's thesis began their essays by paying homage to this "great" and "distinguished" scientist whose work ranks him as "one of the most important intellectuals of our time."[16] Wilson's scholarship in evolutionary biology marks him as a leader in the natural sciences. In addition, his two Pulitzer prizes distinguish him as a person who can draw the ear of a broad audience. As a well-recognized and highly successful Harvard scientist, he has the kind of ethos necessary to make a persuasive claim about what would be the most successful professional moves for ambitious readers to make in the future.

Because of his undisputed professional status, the persona that Wilson developed in his book was closer to Schrödinger's than to Dobzhansky's. Wilson wrote in the informal voice of a scientific popularizer. Like Schrödinger, he often spoke in the first person, referring to events from his past and introducing his own beliefs and hopes. He included many citations to back his claims; however, these citations appeared in a chapter-by-chapter summary at the end of the book, with nothing in the body of the text to mark their existence. This particular citation system allowed those who were so inclined to look up sources, but gave the reader encountering claims no ongoing reassurance that the author was backed by a larger scientific community. In a prototypical scientific "truth-claim" text, such a system would be unacceptable, but in the interdisciplinary inspirational genre, it was quite sufficient, especially since Wilson was a respected scientist whose competence could be trusted.

It was the third characteristic of the genre, that the text be designed to appeal to more than one disciplinary audience, that Wilson's book did not successfully exhibit. Wilson did not negotiate the competing interests of two or more groups; instead, he favored one side over the other. Rather than introduce metaphors that would allow two communities to crisscross their conceptual commitments and thus rethink their intellectual worlds in a way that brought them closer to their opponents, Wilson introduced metaphors of conquest that pitted one side against the other. Rather than encourage polysemous readings that would cause different audience to accept his call to action for different reasons, Wilson's statements were too blatant, extreme, and biased to be interpreted as friendly to both sides of the battle. Unlike Schrödinger, who chose to employ strategic ambiguity to draw both reductionists and nonreductionists to his cause, Wilson supported reductionism in no uncertain terms. And unlike Dobzhansky, who

16. Elshtain, review of *Consilience*, 59; Dupré, "Unification Not Proved," 1395; D. Jamieson, "Cheerleading for Science," 90. See also Jones, "In the Genetic Workshop," 15; Schroeder, "Material Guy," 49; Barr, "Mindless Science," 31; Dyson, review of *Consilience*, 205; Berwick, "All Together Now," 12.

chose to follow his conspicuous division of fields with a subtle but pervasive reunification to reassure scholars from the less powerful side that their methods would be preserved in the coming merger, Wilson favored the more powerful field throughout his book. His few concessions to the humanities and social sciences were so meager and superficial when compared to his rejection of those fields that his treatment could not be viewed as a well-designed Aesopian appeal but only as a clumsy attempt to equivocate. In short, although there is plenty of evidence that Wilson understood the need to address both natural scientists and scholars in the social sciences and humanities, he did not design a text that could effectively do that.[17]

Consilience was meant to serve the social function of an interdisciplinary inspirational monograph, and in some ways, it was well designed to meet the requirements for success in that genre. In analyzing the interdisciplinary space, it summarized the literature that supported a move across disciplinary borders. In addition, it was written by a scholar with the requisite status to gain a hearing on the issue. But because it did not employ effective rhetorical strategies for negotiating the interests of multiple audiences, it was unable to develop a very successful appeal for interdisciplinarity. Ultimately, Wilson was unable to fully achieve his goal because he did not design his text to meet the requirements of the situation.

17. For evidence that Wilson recognized the need to speak to two audiences, see my historical analysis at the beginning of chapter 6.

9

Contributions to Four Ongoing Conversations

Now that we have a better idea of what the genre of interdisciplinary inspiration entails, I would like to draw some conclusions about my study of this genre. In addition to scrutinizing texts that sought to cross disciplinary boundaries, this book has also been designed to *speak* across disciplinary boundaries. Throughout this book, I have sought to balance the interests of at least four different audiences: those who produce scholarship on the rhetoric of science, those who do work in the larger discipline of rhetorical inquiry, those who are interested in the history of science, and those who are interested in how interdisciplinary research communities are initiated. In this chapter, I offer some final comments to each of these audiences.

Rhetoric of Science

To the rhetorician of science, I offer this study as a map of one small region in a territory of scholarship left undeveloped by our subdiscipline's twenty-year focus on epistemological issues. As Charles Alan Taylor argues, our critical study has for too long been restricted by a focus on the hidden rhetorics in knowledge claims and the (de)legitimization of those knowledge claims. It is time for the subdiscipline to broaden the range of what we consider to be rhetorics of science and "move beyond the traditional exemplary texts of science, for example, journal articles or 'revolutionary' documents such as Darwin's *Origin.*" By doing so, "we can begin to come to grips with science as a complex network of cultural practices rather than simply laboratory practices and the claims issued from them."[1] Taylor believes that this sort of study is good for the subdiscipline because it has the potential to improve our understanding of science. In addition, I believe it is of benefit because it has the potential to make our

1. Taylor, "Science as Cultural Practice," 69.

work more relevant to the larger community of rhetoricians of which we are a part.

Building a Better Understanding of Science

At one time, it was a radical position to say that scientific knowledge was shaped through the persuasive arguments of scientific rhetors. So when the founders of the "rhetoric of science" movement began their struggle against received opinion, they focused on the hardest cases. At that time, it did not make sense to waste resources studying the more obviously rhetorical texts of the scientific community. In fact, at that time, a decision to study anything other than the prototypical scientific knowledge claim implied an abandonment of the strong case that the heart of science is open to rhetorical analysis. But times have changed. Because the pioneers of our subdiscipline did their job so well, the idea of a rhetoric of science no longer seems strange to most people. In fact, a focus today on only the "hard case" of the scientific truth claim does more to reify science as a purely cognitive enterprise than it does to open science to the illumination of rhetorical scrutiny.

When we recognize this and expand the range of objects treated by the rhetoric of science community, the main benefit to our understanding of science comes from the more complete picture we develop of how science operates. By acknowledging the importance of the scientific community's internal political negotiations about issues of recruitment, funding, and professional status, we recognize that the development of knowledge is dependent on a complex interaction of cognitive and social factors. Without the inspirational discourse of a scientist such as Schrödinger, molecular biology may not have developed in the way it did and at the time it did. Likewise, without the interdisciplinary rhetoric of a scientist such as Dobzhansky, our modern understanding of evolutionary theory might be quite different. And if Wilson could be more successful at inspiring interdisciplinary activity, the future of the university might shift significantly from the trajectory on which it is currently moving. Studying texts like theirs allows the rhetorician of science to appreciate an important, if indirect, way in which rhetorical considerations define the shape of modern scientific knowledge.

By examining nonprototypical scientific texts, we might also begin to recognize rhetorical strategies that we otherwise would fail to associate with modern science. For example, some theorists who focus solely on the modern prototype of the scientific truth claim believe that scientific writing is designed to eradicate polysemy; according to them, successful scientific texts achieve exegetical equality.[2] Against this premise, my book has argued that some successful twentieth-century scientific texts achieved

2. Markus, "Why Is There No Hermeneutics of Natural Sciences?" 9; McGuire and Melia, "Some Cautionary Strictures," 96; Ricoeur, "Creativity in Language," 127–28.

their effects *because* they were designed to be interpreted in different ways by different audiences. In fact, exegetical equality may work against the social action of the interdisciplinary inspirational monograph of science. This is a discovery about rhetoric in modern science that might not have been made if the subdiscipline were restricted to the study of just one genre of scientific text.[3]

Making Connections to the Broader Discipline of Rhetorical Inquiry

Another reason to expand the range of objects studied by rhetoricians of science is to strengthen the power of the subdiscipline in the larger community of "rhetorical inquiry." Some rhetoricians are still skeptical about how well a disciplinary terminology generated in classical political contexts is able to be transferred to the critique of discourse in the technical sphere of science.[4] Connected to this is a doubt about how well the conclusions of the rhetorical study of scientific texts can be transferred back to aid the continued study of discourse in the public sphere. By examining nonprototypical genres of scientific texts, we can begin to overcome these concerns. Because the genre of interdisciplinary inspirational discourse is about a call to action, not the adjudication of scientific truth, texts in this genre are more obviously open to examination by the vocabulary of the rhetorical tradition, and findings from such a study can be more easily transferred back to the larger community of rhetoricians. For example, the discourse that I've discovered in this genre shows strong analogic connections to the discourse of political negotiation in the civic arena.

Rather than maintain the isolation of a subfield that can speak about the specialized rhetorics of science, but little else, a focus on the more political texts of science allows rhetoricians of science to build bridges back to the larger community and contribute the "theoretical constructs" that critics such as Dilip Gaonkar believe have not been forthcoming from our work.[5] For example, this book's discoveries about how strategies such as conceptual chiasmus and polysemous constructions worked to negotiate collaborative action from conflicting disciplinary groups could be applied by those who study civic discourse in order to better understand negotiations between conflicting political groups in the public sphere.

Rhetorical Inquiry

In producing scholarship on important theoretical constructs that operate both inside and outside the world of science, such as conceptual chiasmus

3. Some critics of Enlightenment and Victorian-era scientific truth claims have recognized that the presence of multiple meanings contributed to the acceptance of those claims. See Campbell, "Polemical Mr. Darwin," 384–85; Campbell, "On the Way," 5, 22–23; and Bishop, "Semantic Flexibility in Scientific Practice," 210–32. My work extends theirs and more directly counters the theorists cited above by showing that *modern* science also occasionally benefits from the presence of polysemy.

4. See for example Gaonkar, "Close Readings," 337–38, 343–44.

5. Gaonkar, "Idea of Rhetoric," 263.

and polysemy, my intent is to contribute not only to the subfield of rhetoric of science but also to the larger field of rhetorical inquiry. The direct application of this book's theoretical insights to the study of nonscientific texts can help make that contribution clear. For example, in an essay that explores the multiple ways in which polysemous constructions work in mass media texts and public address, I used what I learned from my study of the texts by Schrödinger and Dobzhansky to produce a new reading of a historic American public speech. I argued that Henry Grady's 1886 oration "The New South" employed strategic ambiguity to persuade two conflicting audiences to accept his call to action.[6] Like the strategic ambiguity of Schrödinger's "other laws of physics" passage, Grady's polysemous speech encouraged each audience group to interpret it as supporting its own interests rather than the interests of the competing group. Although I have not yet applied this book's findings regarding conceptual chiasmus to other forms of discourse, I suspect that this too is a rhetorical strategy that exists not only in the texts I have examined here, but in a variety of nonscientific texts as well. Further study of this strategy in various contexts could be useful to the field of rhetorical inquiry.

In addition to uncovering theoretical constructs that warrant further study by rhetoricians, this book has introduced a new way of doing rhetorical criticism. This new approach could be the most significant contribution this book makes to the field.

A New Critical Practice

In each case study, a historical reading of conflict between two or more intellectual communities was combined with a close reading of the primary text to reveal specific rhetorical strategies and their probable effects. My findings were then tested against a close intertextual reading of responses written by members of those communities to see whether the rhetorical strategies worked in the way I had supposed. A hypothesis made from my reading of the text was abandoned or modified if a reading of the intertext could not support it. The accuracy of my reading was thus augmented by the constraints that text, historical context, and intertext offered as they were read against one another. The triple redundancy of this new critical practice added a level of rigor and precision that is lacking in the more standard close reading method of rhetorical criticism.

Because this critical practice refuses to refract an interpretation of the text through an appreciative understanding of the "real" meaning of the author or a singular focus on the reader that was conspicuously constructed within the text, it is closely related to the new focus on audience research being taken by critics who study popular cultural texts. For example, Justin Lewis identified ambiguities in an episode of *The Cosby Show* by interviewing different audience groups who had gathered to

6. Ceccarelli, "Polysemy," 405–7.

watch it, and Janice Radway's ethnographic study of women reading Harlequin romance novels established audience interpretations that were different from those offered by literary critics.[7] Audience research has even been applied to the rhetoric of science. For example, Danette Paul and Davida Charney have conducted experiments in which scientists were observed reading scientific texts aloud and then were interviewed about their reading practices.[8]

Although my approach is allied with these audience studies, it differs from most audience research in one very important way. Rather than require the rhetorician to employ interviews, questionnaires, or experimental methods to determine audience response, I urge the rhetorician to apply her method of close textual analysis to the historical traces of reception that were left in such intertextual materials as diaries, letters, editorials, book reviews, and articles that cite the text. Because my approach does not require the rhetorical critic to adopt new, unfamiliar methods, it is more likely to be used by rhetorical critics seeking to improve their studies.

In addition to sharing certain affinities with the new movement toward audience research, my modified critical practice is related to the work of some contemporary literary critics. For example, the recognition that different audiences have different reading frames is nicely captured by Stanley Fish's notion of the "interpretive community," and the practice of tracing reception by closely examining historical evidence that preserves the responses of readers has been demonstrated nicely by Steven Mailloux.[9] My approach has certain hereditary similarities with both of these, but it also differs from them in significant ways. Stanley Fish first introduced the idea of the "interpretive community" to help students recognize that there is no single objective meaning of a text, but neither is meaning wholly subjective. Instead, individuals are constrained by the social systems of intelligibility, the practices and assumptions, that inform the institutional communities in which they are located.[10] He wrote:

> Indeed, it is interpretive communities, rather than either the text or the reader, that produce meanings. . . . An interpretive community is not objective because as a bundle of interests, of particular purposes and goals, its perspective is interested rather than neutral; but by the very same reasoning, the meanings and texts produced by an interpretive community are not subjective because they do not proceed from an isolated individual but from a public and conventional point of view.[11]

7. See Lewis, *Ideological Octopus;* Radway, *Reading the Romance.*

8. Paul and Charney, "Introducing Chaos (Theory)," 396–438; Charney, "Study in Rhetorical Reading," 203–31.

9. Fish, *Is There a Text in This Class?;* Mailloux, *Rhetorical Power.*

10. Fish, *Is There a Text in This Class?* 306, 335.

11. Ibid., 14.

My approach is similar to Fish's in that it recognizes that different audiences produce different meanings for texts because they approach those texts with different sets of assumptions and interests. But our approaches diverge in the purpose toward which each of us applies the concept of interpretive communities. When Fish first proposed the idea, he did so to solve a problem with the reading of literary texts in a classroom environment. By allowing for the possibility of different interpretive communities, the professor could account for divergent but equally plausible readings of a text; and by situating reading strategies in communities of consensual meaning, the professor could limit the range of readings in an environment where institutional constraints required some guidelines on a potentially boundless deconstructive process. As Fish explained,

> This meant that the business of criticism was not (as I had previously thought) to determine a correct way of reading but to determine from which of a number of possible perspectives reading will proceed. This determination will not be made once and for all by a neutral mechanism of adjudication, but will be made and remade again whenever the interests and tacitly understood goals of one interpretive community replace or dislodge the interests and goals of another.[12]

In other words, the function of Fish's method of criticism was to select a single defensible, although temporary, reading of a text from a variety of possible readings. In contrast, I have used the idea of interpretive communities and their different reading frames to illustrate how multiple historically situated groups read a text differently at a particular moment in time. Rather than use the concept as a practical theory for authorizing a single reading of a text, I have used it as a historiographic tool to uncover and organize the multiplicity of readings that existed in the responses to a text that were recorded and preserved in the documents that cited it.[13]

Because I use the concept of interpretive communities as a historiographic tool, my approach is somewhat closer to the one applied by Steven Mailloux when he offered a close reading of responses to Mark Twain's *Huckleberry Finn.* After offering "the interpretation of a hypothetical

12. Ibid., 16.

13. The pedagogical origin of Fish's approach to criticism has led to a number of complaints about its political desirability. See, e.g., Pratt, "Interpretive Strategies/Strategic Interpretations," 224–28; Kent, "Hermeneutical Terror," 132. Both argue that Fish ignores the power relations that exist in the literary criticism classroom; he promotes a theory that serves to validate only one interpretive community at a particular moment in time, while repressing the voice of others. My own approach is not open to the same critique because I have not used the concept of interpretive community to validate a single (if temporary) reading of a text but to reveal multiple valid and competing interpretations at a particular moment in time. In fact, because I seek to authorize multiple contemporaneous meanings, my approach can work to reveal power relations, rather than gloss them, and validate minority voices, rather than silence them.

1880s reading experience described from the perspective of the 1980s," Mailloux decided to "ask questions about how the novel was actually received in the 1880s."[14] He found that, surprisingly, the actual reception differed a great deal from the hypothetical reading experience that today's literary critics imagine when they consider the text in its context. Whereas today's critics interpret the text "within a specific context of political debate over racism after the end of Reconstruction," audience members at the time interpreted the text in the context of a political debate over literary censorship and juvenile delinquency.[15] In demonstrating this difference between modern and historical reception of a literary text, Mailloux was able to show that interpretation is itself a rhetorical process, dependent on the different cultural attitudes brought to bear in different historical periods.

Mailloux's approach is similar to mine in that it moves comfortably between text, historical context, *and* intertextual evidence of reception. It is different from mine in that it traces differences across time, rather than revealing the presence of different interpretive communities at a particular moment in history.[16] In a way, Mailloux's approach is similar to Fish's, since both authorize only one meaning (even if temporary) at any single moment in time.[17]

In addition, our approaches differ in the purpose toward which we put the close reading of receptional evidence. Mailloux is interested in demonstrating that interpretation is politicized and historicized; he wants to show that literary hermeneutics is a rhetorical phenomenon, guided by cultural politics and grounded in acts of persuasion.[18] That is why he is sensitive to a shift in the terms used to critique Twain's book in a literary institution that was engaged in different conversations at different points in history. In contrast, I examine textual reception in order to make a connection between rhetorical strategies intrinsic to the text and their purported extrinsic effects. As a rhetorical scholar (not a literary critic), I am interested in how persuasive discourse works (rather than in how literary theories are validated).[19] That is why I am sensitive to differences in how

14. Mailloux, *Rhetorical Power*, xi.

15. Ibid.

16. Some rhetorical critics in the field of speech communication have, like Mailloux, traced intertextual evidence of reception to show how texts have been reinterpreted *over time*. See M. S. Watson, "Dynamics of Intertextuality," 91–111; Leff, "Lincoln," 131–55.

17. I should note that although I validate multiple meanings at any one point in time, mine is not a deconstructive project that seeks to reduce a text to indeterminate meaning. Instead, I locate multiple determinate meanings in the intersection between textual strategies and interpretive communities.

18. Mailloux, *Rhetorical Power*, 15, 17, 180–81.

19. Michael C. Leff made this second difference clear to me in his conference paper on my dissertation research at the 1997 National Communication Association Convention in Chicago.

the terms of a call to action were received by different groups of people in an immediate audience.

Benefits and Limitations of This New Critical Practice

One benefit of my new method of criticism is its openness to the possibility of multiple meanings. Whereas other methods try to authorize a single meaning for a text, either by locating it in an implied author or a dominant audience, my method recognizes that different interpretations exist, and they are not only valid, but may be necessary to the social action of a text. By documenting the way in which different readings of a text worked together to motivate a single outcome, I am able to explain why some texts were so popular with multiple audiences. Also, by identifying and validating different interpretations, I am able to give voice to readings that may be suppressed by other critical approaches. For example, it would have been difficult, if not impossible, to recognize Dobzhansky's implicit support for naturalists in his first chapter had I produced a close textual criticism that read the text solely from the perspective of the "implied author" and the "ideal audience" as they were constructed in the text. Nor could I have recognized Dobzhansky's subtle move to support naturalists' interests had I followed Fish and Mailloux and examined the text from the reading frame of the geneticists' interpretive community in an attempt to justify a single, temporary, but culturally dominant interpretation at a particular moment in history. But by considering the response of the audience rather than focusing exclusively on the text, and by refusing to validate the reading of only one interpretive community at a time, I was able to recognize the way in which the text's explicit support for a dominant audience was paralleled by subtextual support for a less powerful group of readers. In other words, I was able to recuperate the reading frame of a marginalized audience and uncover the textual evidence that spoke to that audience.[20]

Another benefit of my unique critical practice is that it allows me to recognize subtleties of rhetorical design and interpretation that I otherwise might have missed. For example, when I first read Schrödinger's text in its historical context, I suspected that disciplinary interests might lead biologists and physicists to develop different interpretations of the "other laws of physics" passage.[21] When I later decided to examine the

20. I recognize that this critical practice may not work in all cases; sometimes there is little or no trace of reception left, and sometimes the reception of marginalized groups is not recorded or preserved. But my experiences in conducting this sort of criticism have convinced me that there is a great deal of historical evidence marking reception available to those who search for it, and the recuperation of readings that do not represent those of the "dominant" or "ideal" audience can complicate the rhetorical criticism of texts in interesting and politically meaningful ways.

21. Ceccarelli, "Masterpiece in a New Genre," *Technical Communication Quarterly*.

intertextual evidence that preserved audience reception, I discovered that the interpretive communities were more complex than I had originally proposed. Commitments to reductionism and antireductionism crossed disciplinary boundaries to shape the reading frames of different audiences. After moving back and forth between text, historical context, *and* intertextual response, I was able to recognize the way in which an individual's place in both a professional trajectory and a political-conceptual field cooperated to produce meaning. Rather than recognize only two reading frames, I now recognize several, corresponding to disciplinary affiliation (biologist or physicist) and philosophical commitment (reductionist or antireductionist).

In short, the new critical practice modeled in this book can help rhetorical critics develop more sophisticated readings of texts. It can also help rhetorical critics develop readings that expose subtextual support for marginalized audiences, and it can help explain the effectiveness of texts addressed to multiple audiences. When intertextual materials marking audience reception exist, a gathering of that material, and a close reading of it, can be quite useful to the rhetorical critic.

Of course, there are limitations to this critical practice as well. Though the close textual-intertextual reading does a good job of exploring the connection between the form of a text and its reception by a culturally situated audience, one limitation of this method is its failure to more fully explore broad cultural influences on the production of a text. In other words, studying both text and reception allows us to say something about the specific influences *of* a text, but says little about the broader cultural influences *on* a text. The problem is not that I have completely neglected the issues of ideology and power that are so important to today's critical cultural theorists. Both are strong components of my analysis of textual *reception.*[22] For example, the ideology of reductionism played an important role in the different interpretations of readers who were motivated by Schrödinger's text, just as this ideology helped to explain why many readers were not motivated by Wilson's; and unequal power relations are the key to understanding the different interpretive positions of readers who were motivated by Dobzhansky's text. But though I concentrated on ideology and power in my readings of *reception,* I did not devote the same attention to an analysis of the ideology and power relations that led to the authorial/cultural *production* of these texts. Nor did I examine broad cultural forces such as race, class, and gender that may have influenced the production of these texts. As a result, I missed an opportunity to say some interesting things about ideology and power. For example, an opposition to the ideology of eugenics was probably a strong factor in the

22. Another way of putting this is to say that in my reading of reception, I've focused on audience/cultural production of texts rather than on authorial/cultural production of texts.

development of Dobzhansky's monograph.[23] Likewise, a feminist reading of Schrödinger's writing has the potential to yield some fascinating insights about the cultural assumptions under which he was writing.[24] And Wilson's *Consilience*, like his earlier *Sociobiology*, could be subjected to a cultural critique that locates the ideological origins of its remarks about gender differences, sexual orientation, and "primitive" thought.

I can offer two justifications for my avoidance of this level of analysis. First, any book must have its limits; had I focused more attention on cultural influences that affected the authors' production of texts, I could not have done as thorough a job of describing the connection between texts, audience culture, and reception. Second, there are many scholars who are currently devoting their attention to cultural criticism, describing the broad social forces that influence the production of texts such as these; there are not as many scholars concentrating on the microprocesses that work between audience and text. I believe that an examination of these microprocesses is important if we are to fully understand the influence of texts on culture and the influence of audiences on texts.

History of Science

This book has urged rhetorical critics to improve their scholarship by adopting the historian's sensitivity to traces of textual reception. At the same time, this book has urged historians and scientists to recognize that a study of the rhetorical form and function of influential scientific texts has something important to contribute to our understanding of how science develops.

Historians of science and scientists have been engaged in ongoing conversations about the influence of Dobzhansky's *Genetics and the Origin of Species*, Schrödinger's *What Is Life?* and Wilson's *Consilience*. A scholar taking a rhetorical perspective joins these conversations with a new point of view that may well mediate some of the conflicts that have arisen about the function, value, and meaning of these texts. All three case studies in this book have shown that a close rhetorical reading of texts and their reception can be an effective way of resolving disputes in the history of science by tying together other explanations at the deeper level of textual design.

With regard to Dobzhansky's text, scholars and scientists have proposed four different explanations for its influence: that it popularized the theory of the mathematical population geneticists, that it popularized the findings of the ecological geneticists, that it provided necessary conceptual links between naturalists and geneticists, and that it appealed to the

23. Beatty, "Dobzhansky and Drift."

24. For instance, when describing the power that could come with a knowledge of the code-script he says it could tell one whether an egg is going to develop into "a fly or a maize plant, a rhododendron, a beetle, a mouse or *a woman*" (*SW*, 20–21, emphasis added).

professional motives of its readers. My rhetorical analysis in this book combined all four explanations by offering the connective of a finely tuned textual reading. By providing a more focused explanation of the specific rhetorical strategies that influenced Dobzhansky's audience, I showed that the text functioned in all four ways, and, more important, I explained *how* it was designed to do so.

With regard to Schrödinger's text, there are two disputes in the vast secondary literature, one surrounding the value of the text, the other surrounding its meaning. The first dispute is between those who say a scientific text should be judged by its cognitive content and those who say that it should be judged by its political influence. My rhetorical approach to the text suggested that neither reading is sufficient on its own; instead, we must recognize how Schrödinger's book worked on both a cognitive and a political level. For instance, Schrödinger's reversal of linguistic practices worked on a political level by assuring each side that the things they were most professionally prepared to study were located across disciplinary lines. But it also worked on a cognitive level; the vitalist language applied to atoms allowed biologists to *think* about atoms in a new way and the mechanist language applied to organisms allowed physicists to *think* about organisms in a new way.[25]

The second dispute about Schrödinger's text regards the meaning of the "other laws" passage. Rather than argue that one reading is more correct than another, I have conducted a rhetorical analysis that shows how all three were supported by the text and how all three were important to the interdisciplinary action that was motivated by the text.[26]

With regard to Wilson's text, the dispute in the secondary literature divides critics of Wilson from his supporters. The former argue that his book was unable to achieve consensus because its main point was wrong; the cultural divide between scientists and scholars of the humanities and social sciences is simply too wide to be successfully bridged, and any attempt to inspire interdisciplinary activity between these groups is doomed to failure. The latter argue that his book was not as successful as he had hoped it would be because many of its readers were too closed-minded to recognize the truth of its main point; political affiliations, ideological orientations, professional interests, and disciplinary dogmas blinded them to the genius behind Wilson's call to action. My rhetorical reading of the text suggested that in a way, both perspectives are right. Wilson was wrong

25. Note that my rhetorical account also offered a cognitive explanation insofar as I described the effect of the text as a rational response to its internal argumentative structure rather than attributing its influence to the personality of its author.

26. Again, a rhetorical analysis shows that political and cognitive explanations are both accurate: professional concerns about how to best win a Nobel Prize, or how to contribute to the new science if you have no proficiency in mathematics, drove individuals in physics and biology respectively; at the same time, intellectual concerns about the adequacy of reductionism as an explanatory scheme also guided their interpretations.

to adopt the particular strategies of persuasion that he did, because the result of those strategies was an audience blinded to the genius of his call to action. Unaware of the lessons learned from a study of successful interdisciplinary inspiration, Wilson designed an appeal that did not negotiate conflicting interests but instead maintained distance between the very communities he wanted to unite.

In short, a close rhetorical reading of the text and its reception is well suited to integrate other explanations of the influence of these texts. Rather than reject one historical account or another, and rather than introduce a new reading that conflicts with all the rest, rhetorical inquiry takes a cue from the theory of interdisciplinarity: mediation of conflict between interpretive communities can be attained through an interdisciplinary practice that integrates the differing conceptual and political commitments of controversialists. By closely examining the microstructures of the texts themselves, and the responses of the audiences who read them, we can come to recognize the different explanations of their function, value, and meaning as partial truths that, when combined at the level of textual influence, unite to produce a more complete picture of these books and their influence on the history of science.

Interdisciplinarity

The final audience for this book has been scientists and scholars who are interested in learning about how alliances across fields are forged. To this audience, I offer the findings of these three case studies as a lesson in how interdisciplinary activity can be motivated. Two types of rhetorical strategies have been uncovered in this book's examination of successful interdisciplinary appeals. One was an integrative mechanism, drawing fields together by helping each side recognize that they are on common ground with their counterparts. Integrative strategies that Dobzhansky and Schrödinger used in their successful appeals for interdisciplinarity included conceptual chiasmus to get each side to think more like the other, prolepsis to counter misconceptions about the other field, and footnote apologies to assure both sides that they share the scientific value of precision.[27]

The other type of strategy employed in successful appeals for interdisciplinarity was a nonintegrative mechanism that worked not to assure the two sides that they were on common ground, but to assure each side that *its own special interests and beliefs* would be served if it were to make a move across the disciplinary border. Schrödinger and Dobzhansky

27. These integrative strategies are consistent with the findings of another rhetorical study conducted on an interdisciplinary inspirational monograph—Debra Journet's study of G. G. Simpson's 1944 *Tempo and Mode in Evolution.* Journet explained how Simpson used standard techniques of integration to develop a "boundary rhetoric" that united paleontological and genetic perspectives. See Journet, "Synthesizing Disciplinary Narratives," 140.

employed strategies of polysemy that allowed different groups to interpret their texts differently, and those different readings motivated these different groups, *for different reasons,* to take the mutual action of moving into an interdisciplinary space.

Steve Fuller has suggested that because rhetoricians study political discourse, they are uniquely able to recognize nonintegrative mechanisms such as this; they understand that collaboration does not require a meeting of minds but only a confluence of behaviors.

> Whereas philosophers since Plato have supposed that communication involves speaker and audience partaking of a common form of thought having its own natural integrity, rhetoricians have taken the more interpenetrative view that any apparent meeting of minds is really an instance of strategically suppressed disagreement that enables an audience to move temporarily in a common direction.[28]

My rhetorical study of interdisciplinarity has suggested that appeals to cross intellectual borders might be understood differently by two or more audiences, and, *because of that strategically suppressed disagreement,* succeed at developing unity of action. Schrödinger's strategically ambiguous passage and Dobzhansky's use of an Aesopian appeal both promoted polysemous readings, initiating interdisciplinary collaboration by way of a nonintegrative mechanism.

To some degree, it is counterintuitive to think that a nonintegrative mechanism such as this can motivate successful interdisciplinary action. If people do not understand things in the same way, how can they possibly work together? In fact, the term *interdisciplinary* is often used interchangeably with the term *integrative,* on the assumption that successful synthesis requires an open discussion of how different fields will be made consistent with each other.[29] Because the discovery of polysemy adds something counterintuitive to our understanding of how interdisciplinarity is motivated, it is also somewhat disturbing. After reflecting on the implications of this strategy, there are three concerns that theorists of interdisciplinarity are likely to raise.

First, such nonintegrative strategies seem short-sighted. Focusing on consensus of action rather than on consensus of understanding, a rhetorician might identify polysemy as an important *first* step in the process of collaboration—namely, getting people into the same room and convincing them to work together on the same project. Polysemy cannot be the final word on interdisciplinary collaboration, however. Presumably, there is a point where consensus of understanding must also occur, where integrative strategies must be employed. With this critique I must wholeheartedly agree. Although polysemy is an important motivator of interdisciplinarity,

28. Fuller, *Philosophy, Rhetoric, and the End of Knowledge,* 35.
29. See, e.g., Klein, *Interdisciplinarity,* 15, 191.

it is probably not a good way to sustain interdisciplinary communities once they have been created. My study has only suggested that polysemy can work as an initiator of interdisciplinarity, not as a mechanism for sustaining collaborative work.

Second, some theorists might be uncomfortable with the concept of polysemy because they think it implies that an irrational force is motivating the merger of the two communities. For example, commenting on my work with Dobzhansky's text, Carol Berkenkotter concluded: "I confess that I am dissatisfied with the solely rhetorical view of 'suppressed disagreement that enables an audience to move temporarily in a common direction' . . . as being the basis of interdisciplinary interpenetration, and feel strongly that interpenetration has a sociocognitive dimension."[30] This discomfort is understandable, but I think it is based on a mistaken belief that a lack of consensus is synonymous with a lack of cognition. The fact that a textual construction is read in a polysemous manner does not mean that individuals are no longer responding on a cognitive level. Different interpretive communities are bringing different conceptual schemata and social interests to bear on the text, and as a result, they read it in different but equally valid ways. They then make reasonable decisions based on this reading, a reading that is both social (driven by their professional interests) and cognitive (driven by their intellectual beliefs). There is no reason to think that once we recognize the importance of polysemy in the motivation of interdisciplinary action we will be forced to abandon sociocognitive explanations.[31]

Third, and perhaps most troubling, my implicit recommendation that scientists should adopt strategies of polysemy to inspire interdisciplinarity may not sit well with those who believe that truthful communication is a precondition for ethical rhetoric and who see strategies of polysemy as deceitful. My reply to this critique is to remind readers that ethical judgments are not as simple as we might sometimes wish them to be. To claim that a *form* of utterance is good or evil, without examining its content or the context in which it was made, is imprudent. There are certainly times when the use of a polysemous textual construction is unethical; likewise, there are times when the use of a polysemous textual construction values the diverse perspectives of unequal social groups and is therefore *more* ethical than a straightforward monosemic message.[32]

As a final comment in a more practical vein to scientists and others who are interested in utilizing polysemy to initiate interdisciplinary activity, I

30. Berkenkotter, "Theoretical Issues," 185.

31. In addition, I have shown that strategies of polysemy work side by side with more integrative strategies. So even if one does not acknowledge the sociocognitive dimension of the former, one should appreciate the fact that my study recognizes sociocognitive dimensions in the development of interdisciplinary communities. I do not claim that polysemy is the only successful strategy used in texts of interdisciplinary inspiration.

32. I provide more arguments for this point in Ceccarelli, "Polysemy."

offer this warning: one cannot simply observe the directive to "be vague" and create a successful appeal to a divided audience. To consciously apply a rhetorical technique such as strategic ambiguity, a text would have to be shaped to appeal to the interpretive tendencies of two or more distinct communities, which means that at least two things are required: a knowledge of the assumptions, professional goals, linguistic practices, and other significant interpretive features of the different groups and the creative skill to appeal to those different sets of assumptions, goals, and practices *simultaneously*. The success of Schrödinger's text and Dobzhansky's text indicates the power of such rhetorical techniques; the controversy surrounding Wilson's text indicates the danger such techniques carry if applied ineffectively. Rather than motivate readers for different reasons to engage in collaborative activity, appeals that try to speak to more than one audience but do not properly balance the interests of those audiences are easily dismissed by readers as a form of equivocation. All three case studies have shown that a carefully balanced negotiation of interests is necessary if one wants to use rhetoric to forge alliances across fields.

Bibliography

Abir-Am, Pnina. "The Discourse of Physical Power and Biological Knowledge in the 1930s: A Reappraisal of the Rockefeller Foundation's 'Policy' in Molecular Biology." *Social Studies of Science* 12 (1982): 341–82.

Alexander, Richard D. "Evolution, Human Behavior, and Determinism." In *PSA 1976*, vol. 2, edited by F. Suppe and P. D. Asquith, 3–21. East Lansing: Philosophy of Science Association, 1977.

Allen, Garland E. *Life Science in the Twentieth Century*. New York: John Wiley and Sons, 1975.

———. "Naturalists and Experimentalists: The Genotype and the Phenotype." In *Studies in the History of Biology*, edited by William Coleman and Camille Limoges, vol. 3, 179–209. Baltimore: Johns Hopkins University Press, 1979.

———. "Theodosius Dobzhansky, the Morgan Lab, and the Breakdown of the Naturalist/Experimentalist Dichotomy, 1927–1947." In *The Evolution of Theodosius Dobzhansky: Essays on His Life and Thought in Russia and America*, edited by Mark B. Adams, 87–98. Princeton: Princeton University Press, 1994.

Alper, Joseph S. "Ethical and Social Implications." In *Sociobiology and Human Nature*, edited by Michael S. Gregory, Anita Silvers, and Diane Sutch, 195–212. San Francisco: Jossey-Bass Publishers, 1978.

Anderson, Walter Truett. "Recycling Sociobiology." Review of *Consilience*, by Edward O. Wilson. *Futures* 31 (February 1999): 101–4.

Auerbach, Charlotte. "Induction of Changes in Genes and Chromosomes." *Cold Spring Harbor Symposium on Quantitative Biology* 16 (1951): 199–213.

Ayala, Francisco J. "'Nothing in Biology Makes Sense Except in the Light of Evolution.' Theodosius Dobzhansky: 1900–1975." *Journal of Heredity* 68 (1977): 3–10.

Ayala, Francisco J., and Walter M. Fitch. "Genetics and the Origin of Species: An Introduction." *Proceedings of the National Academy of Sciences of the United States of America* 94 (1997): 7691–97.

Baker, H. G. "The Ecospecies—Prelude to Discussion." *Evolution* 6 (1952): 61–68.

Barkow, Jerome H. "Sociobiology: Is This the New Theory of Human Nature?" In *Sociobiology Examined,* edited by Ashley Montagu, 171–97. New York: Oxford University Press, 1980.

Barnett, S. A. "Biological Determinism and the Tasmanian Native Hen." In *Sociobiology Examined,* edited by Ashley Montagu, 135–57. New York: Oxford University Press, 1980.

Barr, Stephen M. "Mindless Science: The Brain and Edward O. Wilson." Review of *Consilience,* by Edward O. Wilson. *Weekly Standard* 3 (6 April 1998): 31–33.

Bateson, William. "Address of the President of the British Association for the Advancement of Science." *Science* 40 (1914): 287–302.

Bazerman, Charles. *Shaping Written Knowledge: The Genre and Activity of the Experimental Article in Science.* Madison: University of Wisconsin Press, 1988.

Beach, Frank A. "Sociobiology and Interspecific Comparisons of Behavior." In *Sociobiology and Human Nature,* edited by Michael S. Gregory, Anita Silvers, and Diane Sutch, 116–35. San Francisco: Jossey-Bass Publishers, 1978.

Beatty, John. "Dobzhansky and Drift: Facts, Values, and Chance in Evolutionary Biology." In *The Probabilistic Revolution.* Vol. 2, *Ideas in the Sciences,* edited by Lorenz Krüger et al., 271–311. Cambridge: MIT Press, Bradford Books, 1987.

Bechtel, William. "Editor's Commentary: Dobzhansky's Contribution to the Evolutionary Synthesis." In *Integrating Scientific Disciplines,* edited by William Bechtel, 137–41. Dordrecht, The Netherlands: Martinus Nijhoff Publishers, 1986.

Beltrán, Enrique. "The Naturalist in America in 1942 ± 75 Years." *American Naturalist* 78 (1944): 544–55.

Berkenkotter, Carol. "Theoretical Issues Surrounding Interdisciplinary Interpenetration." *Social Epistemology* 9 (1995): 175–87.

Bernstein, Jeremy. "E. O. Wilson's Theory of Everything." Review of *Consilience,* by Edward O. Wilson. *Commentary* 105 (June 1998): 62–65.

Berry, Brian J. L. "On Consilience." *Urban Geography* 19 (1998): 95–97.

Berry, Wendell. *Life Is a Miracle. An Essay against Modern Superstition.* Washington, D.C.: Counterpoint, 2000.

Berwick, Robert C. "All Together Now." Review of *Consilience,* by Edward O. Wilson. *Los Angeles Times Book Review,* 30 August 1998, 12.

Bishop, Michael. "Semantic Flexibility in Scientific Practice: A Study of Newton's Optics." *Philosophy and Rhetoric* 32 (1999): 210–32.

Blackwelder, R. E., and Alan Boyden. "The Nature of Systematics." *Systematic Zoology* 1 (1952): 26–33.

Blair, G. W. S., and B. C. Veinoglou. "Limitations of the Newtonian Time Scale in Relation to Non-equilibrium Rheological States and a Theory of Quasi-Properties." *Proceedings of the Royal Society of London, Series A: Mathematical and Physical Sciences* 189 (1947): 69–87.

Bohr, Niels. "Light and Life." In *Niels Bohr: A Centenary Volume,* edited by A. P. French and P. J. Kennedy, 311–19. Cambridge: Harvard University Press, 1985. First published in *Nature* 131 (1933): 421–53, 457–59.

Boltzmann, Ludwig. "On the Necessity of Atomic Theories in Physics." [1897] In *The Question of the Atom,* edited by M. J. Nye, 357–71. Los Angeles: Tomash, 1984.

Bonner, John Tyler. "A New Synthesis of the Principles That Underlie All Animal Societies." *Scientific American* 233 (October 1975): 129–30, 132.

Boulding, Kenneth E. "Sociobiology or Biosociology?" In *Sociobiology and Human Nature,* edited by Michael S. Gregory, Anita Silvers, and Diane Sutch, 260–76. San Francisco: Jossey-Bass Publishers, 1978.

Boyden, Alan. "Comparative Evolution with Special Reference to Primitive Mechanism." *Evolution* 7 (1953): 21–30.

———. "The Significance of Asexual Reproduction." *Systematic Zoology* 3 (1954): 26–37, 47.

Brandon, Robert. "Introduction: Dobzhansky's Contribution to the Evolutionary Synthesis." In *Integrating Scientific Disciplines,* edited by William Bechtel, 109–11. Dordrecht, The Netherlands: Martinus Nijhoff Publishers, 1986.

Brierley, William B. Review of *Genetics and the Origin of Species,* by Theodosius Dobzhansky. *Annals of Applied Biology* 25 (1938): 667–69.

Brillouin, L. "Life, Thermodynamics, and Cybernetics." *American Scientist* 37 (1949): 554–68.

Bunk, Steve. "Is Science Religious? Why Do These Often Opposing Pursuits Engender Similar Emotions?" *Scientist* 13, no. 22 (1999): 10–11.

Burian, Richard M. "Challenges to the Evolutionary Synthesis." *Evolutionary Biology* 23 (1988): 247–69.

———. "The Influence of the Evolutionary Paradigm." In *Evolutionary Biology at the Crossroads: a Symposium at Queens College,* edited by Max K. Hecht, 149–66. Flushing, N.Y.: Queens College Press, 1989.

Burnett, D. Graham. "A Dream of Reason." Review of *Consilience,* by Edward O. Wilson. *American Scholar* 67 (summer 1998): 143–47.

———. "A View from the Bridge: The Two Cultures Debate, Its Legacy, and the History of Science." *Daedalus* 128 (spring 1999): 193–218.

Butler, J. A. V. "Life and the Second Law of Thermodynamics." *Nature* 158 (1946): 153–54.

Cain, Joseph Allen. "Common Problems and Cooperative Solutions: Organizational Activity in Evolutionary Studies, 1936–1947." *Isis* 84 (1993): 1–25.

Camerini, Jane R. "Evolution, Biogeography, and Maps: An Early History of Wallace's Line." *Isis* 84 (1993): 700–727.

Campbell, John Angus. "The Polemical Mr. Darwin." *Quarterly Journal of Speech* 61 (1975): 375–90.

———. "On the Way to the Origin: Darwin's Evolutionary Insight and Its Rhetorical Transformation." *Van Zelst Lecture in Communication.* Evanston: Northwestern University, 1990.

Campbell, John Angus, and Keith Benson. "The Rhetorical Turn in Science Studies." *Quarterly Journal of Speech* 82 (1996): 74–109.

Carlson, Elof Axel. *The Gene: A Critical History.* Philadelphia: W. B. Saunders, 1966.

———. "The Unacknowledged Founding of Molecular Biology: H. J. Muller's Contributions to Gene Theory, 1910–1936." *Journal of the History of Biology* 4 (1971): 149–70.

———. "H. J. Muller: The Role of the Scientist in Creating and Applying Knowledge." *Social Research* 51 (1984): 763–82.

Carman, James M. Review of *Consilience,* by Edward O. Wilson. *Journal of Macromarketing* 18 (fall 1998):169–71.

Carson, Hampton L. "Cytogenetics and the Neo-Darwinian Synthesis." In *The Evolutionary Synthesis: Perspectives on the Unification of Biology*, edited by Ernst Mayr and William B. Provine, 86–95. Cambridge: Harvard University Press, 1980.

———. "Dobzhansky and the Deepening of Darwinism." *Genetics of Natural Populations: The Continuing Importance of Theodosius Dobzhansky*, edited by Louis Levine, 56–70. New York: Columbia University Press, 1995.

Ceccarelli, Leah. "A Masterpiece in a New Genre: The Rhetorical Negotiation of Two Audiences in Erwin Schrödinger's *What Is Life?*" Master's thesis, Northwestern University, 1992.

———. "A Masterpiece in a New Genre: The Rhetorical Negotiation of Two Audiences in Schrödinger's *What Is Life?*" *Technical Communication Quarterly* 3 (1994): 7–17.

———. "Polysemy: Multiple Meanings in Rhetorical Criticism." *Quarterly Journal of Speech* 84 (November 1998): 395–415.

Chargaff, Erwin. *Heraclitean Fire: Sketches from a Life before Nature.* New York: Rockefeller University Press, 1978.

Charney, Davida. "A Study in Rhetorical Reading: How Evolutionists Read 'The Spandrels of San Marco.'" In *Understanding Scientific Prose*, edited by Jack Selzer, 203–31. Madison: University of Wisconsin Press, 1993.

Cohen, Seymour S. "The Origins of Molecular Biology." *Science* 187 (1975): 827–30.

Cole, Leon J. "Each After His Kind." *Science* 93 (1941): 289–93, 316–19.

Condit, Celeste Michelle. "Rhetorical Criticism and the Audience: The Extremes of McGee and Leff." *Western Journal of Speech Communication* 54 (1990): 330–45.

"Consilience, Complexity, and Communication: Three Challenges at the Start of the New Century." *BioEssays* 21 (1999): 983–84.

Costanza, Robert. "One Giant Leap." Review of *Consilience*, by Edward O. Wilson. *BioScience* 49 (June 1999): 487–88.

Crick, F. H. C. "Recent Research in Molecular Biology: Introduction." *British Medical Bulletin* 21 (1965): 183–86.

Darden, Lindley. "Relations among Fields in the Evolutionary Synthesis." In *Integrating Scientific Disciplines*, edited by William Bechtel, 113–23. Dordrecht, The Netherlands: Martinus Nijhoff Publishers, 1986.

Darlington, C. D. Review of *What Is Life?* by Erwin Schrödinger. *Discovery* 6 (1945): 126.

Delbrück, Max. "What Is Life? and What Is Truth?" Review of *What Is Life?* by Erwin Schrödinger. *Quarterly Review of Biology* 20 (1945): 370–72.

———. "A Physicist Looks at Biology." In *Phage and the Origin of Molecular Biology: Expanded Edition*, edited by John Cairns, Gunther S. Stent, and James D. Watson, 9–22. Cold Spring Harbor: Cold Spring Harbor Laboratory Press, 1992. First published in *Transactions of the Connecticut Academy of Arts and Sciences* 38 (1949): 173–90.

Depew, David J., and Bruce H. Weber. "Innovation and Tradition in Evolutionary Theory: An Interpretive Afterword." In *Evolution at a Crossroads: The New Biology and the New Philosophy of Science*, edited by David J. Depew and Bruce H. Weber, 227–60. Cambridge: MIT Press, 1985.

———. "Consequences of Nonequilibrium Thermodynamics for the Darwinian

Tradition." In *Entropy, Information, and Evolution: Perspectives on Physical and Biological Evolution,* edited by Bruce H. Weber, David J. Depew, and James D. Smith, 317–42. Cambridge: MIT Press, 1988.

———. *Darwinism Evolving: Systems Dynamics and the Genealogy of Natural Selection.* Cambridge: MIT Press, 1995.

Dobzhansky, Theodosius. *Genetics and the Origin of Species.* New York: Columbia University Press, 1937; reprint, New York: Columbia University Press, 1982.

———. *Genetics and the Origin of Species.* 2d ed. New York: Columbia University Press, 1941.

———. *Genetics and the Origin of Species.* 3d ed. New York: Columbia University Press, 1951.

Doyle, Michael P. "Biology Takes Center Stage." Review of *Consilience,* by Edward O. Wilson. *Chemical and Engineering News* 76 (12 October 1998): 64–65.

Driesch, Hans. "The Breakdown of Materialism." In *The Great Design: Order and Intelligence in Nature,* edited by F. Mason, 2:283–303. New York: Macmillan, 1935.

Dunn, L. C. Foreword to *Genetics and the Origin of Species,* by Theodosius Dobzhansky. New York: Columbia University Press, 1937.

Dupré, John. "Unification Not Proved." Review of *Consilience,* by Edward O. Wilson. *Science* 280 (1998): 1395.

Dyson, Freeman. Review of *Consilience,* by Edward O. Wilson. *New England Journal of Medicine* 339 (16 July 1998): 205.

Eldredge, Niles. *Unfinished Synthesis: Biological Hierarchies and Modern Evolutionary Thought.* Oxford: Oxford University Press, 1985.

———. "Cornets and Consilience." *Civilization* 5 (October–November 1998): 84–86.

Elitzur, Avshalom C. "Life and Mind, Past and Future: Schrödinger's Vision Fifty Years Later." *Perspectives in Biology and Medicine* 38 (1995): 433–58.

Ellis, George E. R. "Nancey Murphy's Work." *Zygon* 34 (1999): 601–7.

Elshtain, Jean Bethke. Review of *Consilience,* by Edward O. Wilson. *First Things: A Journal of Religion and Public Life* 91 (March 1999): 59.

Emerson, Alfred E. "The Origin of Species." Review of *Genetics and the Origin of Species,* by Theodosius Dobzhansky. *Ecology* 19 (1938): 152–54.

Fischer, Ernst Peter. "We Are All Aspects of One Single Being: An Introduction to Erwin Schrödinger." *Social Research* 51 (1984): 809–35.

Fish, Stanley. *Is There a Text in This Class? The Authority of Interpretive Communities.* Cambridge: Harvard University Press, 1980.

Fisher, George Ross. Review of *Consilience,* by Edward O. Wilson. *JAMA* 280 (28 October 1998): 1455.

Fisher, Walter R. "Narration, Knowledge, and the Possibility of Wisdom." In *Rethinking Knowledge: Reflections across the Disciplines,* edited by Robert F. Goodman and Walter R. Fisher, 169–92. New York: State University of New York Press, 1995.

Fleming, Donald. "Émigré Physicists and the Biological Revolution." In *Perspectives in American History* 2:152–89. Cambridge: Harvard University Press, 1968.

Forehand, Rex. "Clinical Child and Developmental-Clinical Programs: Perhaps

Necessary But Not Sufficient?" *Journal of Clinical Child Psychology* 28 (1999): 476–84.

Forrest, David V. Review of *Consilience*, by Edward O. Wilson. *American Journal of Psychiatry* 155 (December 1998): 1796–97.

Freeman, Derek. "Sociobiology: The 'Antidiscipline' of Anthropology." In *Sociobiology Examined*, edited by Ashley Montagu, 198–219. New York: Oxford University Press, 1980.

Fuerst, John A. "The Role of Reductionism in the Development of Molecular Biology: Peripheral or Central." *Social Studies of Science* 12 (1982): 241–78.

Fuller, Steve. *Philosophy, Rhetoric, and the End of Knowledge: The Coming of Science and Technology Studies.* Madison: University of Wisconsin Press, 1993.

Gaonkar, Dilip Parameshwar. "The Idea of Rhetoric in the Rhetoric of Science." *Southern Speech Communication Journal* 58 (1993): 258–95.

———. "Close Readings of the Third Kind: Reply to My Critics." In *Rhetorical Hermeneutics: Invention and Interpretation in the Age of Science*, edited by Alan G. Gross and William M. Keith, 330–56. Albany: State University of New York Press, 1996.

Gerstel, D. U. "A New Lethal Combination in Interspecific Cotton Hybrids." *Genetics* 39 (1954): 628–39.

Gillispie, Charles C. "E. O. Wilson's *Consilience:* A Noble, Unifying Vision, Grandly Expressed." Review of *Consilience*, by Edward O. Wilson. *American Scientist* 86 (May–June 1998): 280–83.

Glass, Bentley. Introduction to *The Roving Naturalist: Travel Letters of Theodosius Dobzhansky*, edited by Bentley Glass. Philadelphia: American Philosophical Society, 1980.

Gould, Stephen Jay. "Sociobiology and Human Nature: A Postpanglossian Vision." In *Sociobiology Examined*, edited by Ashley Montagu, 283–90. New York: Oxford University Press, 1980.

———. Introduction to *Genetics and the Origin of Species*, by Theodosius Dobzhansky. New York: Columbia University Press, 1982.

———. "In Gratuitous Battle." *Civilization* 5 (October–November 1998): 86–88.

Gregory, Michael S. "Epilogue." In *Sociobiology and Human Nature*, edited by Michael S. Gregory, Anita Silvers, and Diane Sutch, 283–94. San Francisco: Jossey-Bass Publishers, 1978.

Gregory, W. K. "Genetics versus Paleontology." *American Naturalist* 51 (1917): 622–35.

Grene, Marjorie. "Sociobiology and the Human Mind." In *Sociobiology and Human Nature*, edited by Michael S. Gregory, Anita Silvers, and Diane Sutch, 213–24. San Francisco: Jossey-Bass Publishers, 1978.

Griffin, Donald R. "Humanistic Aspects of Ethology." In *Sociobiology and Human Nature*, edited by Michael S. Gregory, Anita Silvers, and Diane Sutch, 240–59. San Francisco: Jossey-Bass Publishers, 1978.

Gross, Alan G. *The Rhetoric of Science.* Cambridge: Harvard University Press, 1990; reprint, Cambridge: Harvard University Press, 1996.

Gross, Paul R. "The Icarian Impulse." *Wilson Quarterly* 22 (winter 1998): 39–49.

Grüneberg, Hans. Review of *Genetics and the Origin of Species*, by Theodosius Dobzhansky. *Eugenics Review* 30 (April 1938): 69–70.

Haldane, J. B. S. "A Physicist Looks at Genetics." Review of *What Is Life?* by Erwin Schrödinger. *Nature* 155 (1945): 375–76.

Hales, Steven D. "The Problem of Intuition." *American Philosophical Quarterly* 37 (2000): 135–47.

Halloran, S. Michael. "The Birth of Molecular Biology: An Essay in the Rhetorical Criticism of Scientific Discourse." *Rhetoric Review* 3 (1984): 70–83.

Hardin, Garrett. "Nice Guys Finish Last." In *Sociobiology and Human Nature,* edited by Michael S. Gregory, Anita Silvers, and Diane Sutch, 183–94. San Francisco: Jossey-Bass Publishers, 1978.

Harris, R. Allen. "Rhetoric of Science." *College English* 53 (1991): 282–307.

Harvey, Paul H. "Further Steps Toward One Culture." Review of *Consilience,* by Edward O. Wilson. *Nature* 392 (1998): 451–52.

Hirsch, Jerry. "The Pitfalls of Heritability: Can All Tangible Phenomena Really Be Reduced to the Laws of Physics?" Review of *Consilience,* by Edward O. Wilson. *Times Literary Supplement,* 12 February 1999, 33.

Holton, Gerald. "The New Synthesis?" In *Sociobiology and Human Nature,* edited by Michael S. Gregory, Anita Silvers, and Diane Sutch, 75–97. San Francisco: Jossey-Bass Publishers, 1978.

Hrdlicka, Ales. "Genetics." Review of *Genetics and the Origin of Species,* by Theodosius Dobzhansky. *American Journal of Physical Anthropology* 24 (1938): 240.

Hubbs, Carl L. Review of *What Is Life?* by Erwin Schrödinger. *The American Naturalist* 79 (1945): 554–55.

Hull, David. "Scientific Bandwagon or Traveling Medicine Show?" In *Sociobiology and Human Nature,* edited by Michael S. Gregory, Anita Silvers, and Diane Sutch, 136–63. San Francisco: Jossey-Bass Publishers, 1978.

Infeld, Leopold. "Visit to Dublin." *Scientific American* 181 (1949): 11–15.

Jacob, François. *The Logic of Life: A History of Heredity.* Translated by Betty E. Spillmann. New York: Pantheon Books, 1973.

Jamieson, Dale. "Cheerleading for Science." Review of *Consilience,* by Edward O. Wilson. *Issues in Science and Technology* 15 (fall 1998): 90–91.

Jamieson, Kathleen Hall. "The Cunning Rhetor, the Complicitous Audience, the Conned Censor, and the Critic." *Communication Monographs* 57 (1990): 73–78.

Jones, Steve. "In the Genetic Toyshop." Review of *Clone: The Road to Dolly and the Path Ahead,* by Gina Kolata, *The Biotech Century: Harnessing the Gene and Remaking the World,* by Jeremy Rifkin, and *Consilience,* by Edward O. Wilson. *New York Review of Books* 45 (23 April 1998): 14–16

Journet, Debra. "Synthesizing Disciplinary Narratives: George Gaylord Simpson's *Tempo and Mode in Evolution.*" *Social Epistemology* 9 (1995): 113–50.

Judson, Horace Freeland. *The Eighth Day of Creation: Makers of the Revolution in Biology.* New York: Simon and Schuster, 1979.

Just, Theodor. Review of *Genetics and the Origin of Species,* by Theodosius Dobzhansky. *American Midland Naturalist* 18 (1937): 1105–6.

Keller, Evelyn Fox. "Physics and the Emergence of Molecular Biology: A History of Cognitive and Political Synergy." *Journal of the History of Biology* 23 (1990): 389–409.

Kendrew, John C. "How Molecular Biology Started." *Scientific American* 216 (1967): 141–44.

Kent, Thomas. "Hermeneutical Terror and the Myth of Interpretive Consensus." *Philosophy and Rhetoric* 25 (1992): 124–39.

Kevles, Daniel J. "The New Enlightenment." Review of *Consilience*, by Edward O. Wilson. *New York Times Book Review* (26 April 1998): 11–12.

Khân, Shâms-ul-Islam. "Pollen Sterility in *Solanum ruberosum L.*" *Cytologia* 16 (1951): 124–30.

Klein, Julie Thompson. *Interdisciplinarity: History, Theory, and Practice.* Detroit: Wayne State University Press, 1990.

Kofoid, Charles A. Review of *Genetics and the Origin of Species*, by Theodosius Dobzhansky. *Isis* 30 (1939): 549–51.

Krimbas, Costas B. "The Evolutionary Worldview of Theodosius Dobzhansky." In *The Evolution of Theodosius Dobzhansky: Essays on His Life and Thought in Russia and America*, edited by Mark B. Adams. Princeton: Princeton University Press, 1994.

Kurtz, Paul. "Can the Sciences Be Unified?" Review of *Consilience*, by Edward O. Wilson. *Skeptical Inquirer* 22 (July–August 1998): 47–49.

Lanier, Jarod. "Biology Rules." *Civilization* 5 (October–November 1998): 83–84.

Lawrence, W. J. C. "Studies on *Streptocarpus.* II. Complementary Sublethal Genes." *Journal of Genetics* 48 (1947): 16–30.

Lawson-Tancred, Hugh. "The Start of an Intellectual Battle." Review of *Consilience*, by Edward O. Wilson. *Spectator* 281 (10 October 1998): 39–40.

Leeds, Anthony. "The Language of Sociobiology: Reduction, Emergence, History, Social Science, Normativeness." *Philosophical Forum* 13 (winter–spring 1981–82): 161–206.

Leff, Michael. "Things Made by Words: Reflections on Textual Criticism." *Quarterly Journal of Speech* 78 (1992): 223–31.

———. "Lincoln Among the Nineteenth-Century Orators." In *Rhetoric and Political Culture in Nineteenth-Century America*, edited by Thomas W. Benson, 131–55. East Lansing: Michigan State University Press, 1997.

Levene, Howard, Lee Ehrman, and Rollin Richmond. "Theodosius Dobzhansky up to Now." In *Essays in Evolution and Genetics in Honor of Theodosius Dobzhansky*, edited by Max K. Hecht and William C. Steere, 1–40. New York: Appleton-Century-Crofts, 1970.

Lewis, Justin. *The Ideological Octopus: An Exploration of Television and Its Audience.* New York: Routledge, 1991.

Lewontin, Richard C. "Sociobiology: A Caricature of Darwinism." In *PSA 1976*, vol. 2, edited by F. Suppe and P. D. Asquith, 22–31. East Lansing: Philosophy of Science Association, 1977.

———. "Theoretical Population Genetics in the Evolutionary Synthesis." In *The Evolutionary Synthesis: Perspectives on the Unification of Biology*, edited by Ernst Mayr and William B. Provine, 58–68. Cambridge: Harvard University Press, 1980.

Lindquist, K. "The Mutant 'Micro' in Pisum." *Hereditas* 37 (1951): 389–420.

Loeb, Jacques. "The Mechanistic Conception of Life." [1911/1912] In *The Mechanistic Conception of Life*, edited by Donald Fleming, 5–34. Cambridge: Belknap Press of Harvard University Press, 1964.

Lumsden Charles J., and Edward O. Wilson. *Genes, Mind, and Culture: The Coevolutionary Process.* Cambridge: Harvard University Press, 1981.

———. *Promethean Fire: Reflections on the Origin of Mind.* Cambridge: Harvard University Press, 1983.

Lyne, John, and Henry F. Howe. "The Rhetoric of Expertise: E. O. Wilson and Sociobiology." *Quarterly Journal of Speech* 76 (1990): 134–51.

Mackintosh, N. J. "A Proffering of Underpinnings." In *Sociobiology Examined,* edited by Ashley Montagu, 336–41. New York: Oxford University Press, 1980.

Mailloux, Steven. *Rhetorical Power.* Ithaca: Cornell University Press, 1989.

Manton, I. "Comments on Chromosome Structure." *Nature* 155 (1945): 471–73.

Markus, Gyorgy. "Why Is There No Hermeneutics of Natural Sciences? Some Preliminary Theses." *Science in Context* 1 (1987): 5–51.

May, Robert. "An Attempt to Link the Sciences and the Humanities." Review of *Consilience,* by Edward O. Wilson. *Scientific American* 278 (June 1998): 97–98.

Mayr, Ernst. Review of *Genetics and the Origin of Species,* by Theodosius Dobzhansky. *Auk* 55 (1938): 300–301.

———. "Speciation Phenomena in Birds." *American Naturalist* 74 (1940): 249–78.

———. "G. G. Simpson." In *The Evolutionary Synthesis: Perspectives on the Unification of Biology,* edited by Ernst Mayr and William B. Provine, 452–63. Cambridge: Harvard University Press, 1980.

———. "How I Became a Darwinian." In *The Evolutionary Synthesis: Perspectives on the Unification of Biology,* edited by Ernst Mayr and William B. Provine, 413–23. Cambridge: Harvard University Press, 1980.

———. "Prologue: Some Thoughts on the History of the Evolutionary Synthesis." In *The Evolutionary Synthesis: Perspectives on the Unification of Biology,* edited by Ernst Mayr and William B. Provine, 1–48. Cambridge: Harvard University Press, 1980.

———. "The Role of Systematics in the Evolutionary Synthesis." In *The Evolutionary Synthesis: Perspectives on the Unification of Biology,* edited by Ernst Mayr and William B. Provine, 123–36. Cambridge: Harvard University Press, 1980.

———. *The Growth of Biological Thought: Diversity, Evolution, and Inheritance.* Cambridge: Belknap Press of Harvard University Press, 1982.

———. "How Biology Differs from the Physical Sciences." In *Evolution at a Crossroads: The New Biology and the New Philosophy of Science,* edited by David J. Depew and Bruce H. Weber, 43–63. Cambridge: MIT Press, 1985.

Mayr, Ernst, and William B. Provine. Preface to *The Evolutionary Synthesis: Perspectives on the Unification of Biology,* edited by Ernst Mayr and William B. Provine, ix–xi. Cambridge: Harvard University Press, 1980.

McElroy, W. D., and C. P. Swanson. "The Theory of Rate Processes and Gene Mutation." *The Quarterly Review of Biology* 26 (1951): 348–63.

McGuire, J. E., and Trevor Melia. "Some Cautionary Strictures on the Writing of the Rhetoric of Science." *Rhetorica* 7 (1989): 87–99.

McShea, Daniel W. "Gene-talk about Sociobiology." *Social Epistemology* 6 (1992): 183–92.

Medawar, P. B. Review of *What Is Life?* by Erwin Schrödinger. In *Biology and Human Affairs* (1945). Reviews of *What Is Life?* Cambridge University Press Archives.

Meglitsch, Paul A. "On the Nature of the Species." *Systematic Zoology* 3 (1954): 49–65.

Midgley, Mary. "Rival Fatalisms: The Hollowness of the Sociobiology Debate." In *Sociobiology Examined*, edited by Ashley Montagu, 13–38. New York: Oxford University Press, 1980.

———. "A Well-Meaning Cannibal." Review of *Consilience*, by Edward O. Wilson. *Commonweal* 125 (July 17, 1998): 23–24.

Miller, Carolyn. "Genre as Social Action." *Quarterly Journal of Speech* 70 (1984): 151–67.

———. "*Kairos* in the Rhetoric of Science." In *A Rhetoric of Doing: Essays on Written Discourse in Honor of James L. Kinneavy*, edited by Stephen P. Witte, Neil Nakadate, and Roger D. Cherry, 310–27. Carbondale: Southern Illinois University Press, 1992.

Moore, John A. "The History of Evolutionary Concepts and the Nature of the Controversies." In *Evolutionary Biology at the Crossroads: A Symposium at Queens College*, edited by Max K. Hecht, 5–19. Flushing, N.Y.: Queens College Press, 1989.

Moore, Walter. *Schrödinger: Life and Thought.* Cambridge: Cambridge University Press, 1989.

Morison, Robert S. "The Biology of Behavior." Review of *Sociobiology: The New Synthesis*, by Edward O. Wilson. *Natural History* (November 1975): 85–89.

Muller, H. J. "Physics in the Attack on the Fundamental Problems of Genetics." *Scientific Monthly* 44 (1936): 210–14.

———. "A Physicist Stands Amazed at Genetics." Review of *What Is Life?* by Erwin Schrödinger. *Journal of Heredity* 37 (March 1946): 90–92.

Mullins, Nicolas C. "The Development of a Scientific Specialty: The Phage Group and the Origins of Molecular Biology." *Minerva* 10 (1972): 51–82.

Myers, Greg. "Every Picture Tells a Story: Illustrations in E. O. Wilson's *Sociobiology*." *Human Studies* 11 (1988): 235–69.

———. *Writing Biology: Texts in the Social Construction of Scientific Knowledge.* Madison: University of Wisconsin Press, 1990.

Olby, Robert. "Schrödinger's Problem: What Is Life?" *Journal of the History of Biology* 4 (1971): 119–48.

———. *The Path to the Double Helix.* Seattle: University of Washington Press, 1974.

Paul, Danette, and Davida Charney. "Introducing Chaos (Theory) into Science and Engineering: Effects of Rhetorical Strategies on Scientific Readers." *Written Communication* 12 (October 1995): 396–438.

Pauling, Linus. "Schrödinger's Contribution to Chemistry and Biology." In *Schrödinger: Centenary Celebration of a Polymath*, edited by C. W. Kilmister, 225–33. Cambridge: Cambridge University Press, 1987.

Pauly, Philip J. *Controlling Life: Jacques Loeb and the Engineering Ideal in Biology.* New York: Oxford University Press, 1987.

Pearl, Raymond. Review of *Genetics and the Origin of Species*, by Theodosius Dobzhansky. *Quarterly Review of Biology* 13 (1938): 211–12.

Perutz, M. F. "Erwin Schrödinger's *What Is Life?* and Molecular Biology." In *Schrödinger: Centenary Celebration of a Polymath*, edited by C. W. Kilmister, 234–51. Cambridge: Cambridge University Press, 1987.

Peter, Karl, and Nicholas Petryszak. "Sociobiology versus Biosociology." In *Socio-*

biology Examined, edited by Ashley Montagu, 39–81. New York: Oxford University Press, 1980.

Pinker, Steven. "The Theory of Everything: E. O. Wilson Explains How All Knowledge Fits Together." Review of *Consilience*, by Edward O. Wilson. *Slate* (31 March 1998) <http://www.slate.com/BookReview/98-03-31/BookReview.asp> (27 July 1999).

Polanyi, M. "Science and Life." Review of *What Is Life?* by Erwin Schrödinger. *The Manchester Guardian*, 26 January 1945, 3.

Pope, Steve. "A Scientist's Search for Comprehensive Knowledge." Review of *Consilience*, by Edward O. Wilson. *Christian Century* 115 (4 November 1998): 1027–33.

Powell, Jeffrey R. "'In the Air'—Theodosius Dobzhansky's *Genetics and the Origin of Species*." *Genetics* 117 (1987): 363–66.

———. "The Neo-Darwinian Legacy for Phylogenetics." In *Genetics of Natural Populations: The Continuing Importance of Theodosius Dobzhansky*, edited by Louis Levine, 71–86. New York: Columbia University Press, 1995.

Pratt, Mary Louise. "Interpretive Strategies/Strategic Interpretations: On Anglo-American Reader Response Criticism." *Boundaries* 2, no. 11 (1982–1983): 224–28.

Prelli, Lawrence. *A Rhetoric of Science: Inventing Scientific Discourse*. Columbia: University of South Carolina Press, 1989.

Provine, William B. "The Role of Mathematical Population Geneticists in the Evolutionary Synthesis of the 1930s and 1940s." In *Studies in the History of Biology*, edited by William Coleman and Camille Limoges, 2:167–92. Baltimore: Johns Hopkins University Press, 1978.

———. "Francis B. Sumner and the Evolutionary Synthesis." In *Studies in the History of Biology*, edited by William Coleman and Camille Limoges, 3:211–40. Baltimore: Johns Hopkins University Press, 1979.

———. "Epilogue." In *The Evolutionary Synthesis: Perspectives on the Unification of Biology*, edited by Ernst Mayr and William B. Provine, 399–411. Cambridge: Harvard University Press, 1980.

———. "Origins of the Genetics of Natural Populations Series." In *Dobzhansky's Genetics of Natural Populations I–XLIII*, edited by R. C. Lewontin et al., 1–85. New York: Columbia University Press, 1981.

———. *Sewall Wright and Evolutionary Biology*. Chicago: University of Chicago Press, 1986.

———. "The Origin of Dobzhansky's *Genetics and the Origin of Species*." In *The Evolution of Theodosius Dobzhansky: Essays on His Life and Thought in Russia and America*, edited by Mark B. Adams, 99–114. Princeton: Princeton University Press, 1994.

Radway, Janice. *Reading the Romance: Women, Patriarchy, and Popular Literature*. Chapel Hill: University of North Carolina Press, 1984.

Reitz, John R., and Conrad Longmire. "Living Matter and Physical Laws." *Physics Today* 3 (1950): 15–19.

Rendel, J. M. "Genetics and Cytology of *Drosophila subobscura*. II. Normal and Selective Matings in *Drosophila subobscura*." *Journal of Genetics* 46 (1945): 287–302.

Review of *What Is Life?* by Erwin Schrödinger. *British Medical Journal* (1945). Reviews of *What Is Life?* Cambridge University Press Archives.

Review of *What Is Life?* by Erwin Schrödinger. In *Plant Breeding Abstracts* 19 (1949): 57. Reviews of *What Is Life?* Cambridge University Press Archives.

Ricoeur, Paul. "Creativity in Language: Word, Polysemy, and Metaphor." In *The Philosophy of Paul Ricoeur: An Anthology of His Work*, edited by Charles E. Reagan and David Stewart, 120–33. Boston: Beacon Press, 1978.

Rorty, Richard. "Against Unity." *Wilson Quarterly* 22 (winter 1998): 28–38.

Rose, Steven. "'It's Only Human Nature': The Sociobiologist's Fairyland." In *Sociobiology Examined*, edited by Ashley Montagu, 158–170. New York: Oxford University Press, 1980.

———. Review of *Consilience*, by Edward O. Wilson. *Endeavour* 23 (March 1999): 36–37.

Rosen, Robert. "The Schrödinger Question: What Is Life? Fifty Years Later." In *Glimpsing Reality: Ideas in Physics and the Link to Biology*, edited by Paul Buckley and F. David Peat, 168–90. University of Toronto Press, 1996.

Ruse, Michael. *Sociobiology: Sense or Nonsense?* Dordrecht, The Netherlands: D. Reidel Publishing, 1979.

Sapp, Jan. "The Struggle for Authority in the Field of Heredity, 1900–1932: New Perspectives on the Rise of Genetics." *Journal of the History of Biology* 16 (1983): 311–42.

Sarkar, Sahotra. "What Is Life? Revisited." *BioScience* 41 (1991): 631–34.

Schmidt, Karl P. Review of *Genetics and the Origin of Species*, by Theodosius Dobzhansky. *Copeia*, 1938 (1938): 51.

Schneewind, J. B. "Sociobiology, Social Policy, and Nirvana." In *Sociobiology and Human Nature*, edited by Michael S. Gregory, Anita Silvers, and Diane Sutch, 225–39. San Francisco: Jossey-Bass Publishers, 1978.

Schneider, Eric D. "Schrödinger's Grand Theme Shortchanged." *Nature* 328 (1987): 300.

Schrödinger, Erwin. *What Is Life? The Physical Aspect of the Living Cell.* Cambridge: Cambridge University Press, 1944.

Schroeder, Gerald L. "Material Guy." Review of *Consilience*, by Edward O. Wilson. *National Review* 50 (4 May 1998): 49–50.

Segerstråle, Ullica. *Defenders of the Truth: The Battle for Science in the Sociobiology Debate and Beyond.* Oxford: Oxford University Press, 2000.

Shapere, Dudley. "The Meaning of the Evolutionary Synthesis." In *The Evolutionary Synthesis: Perspectives on the Unification of Biology*, edited by Ernst Mayr and William B. Provine, 388–98. Cambridge: Harvard University Press, 1980.

Simon, Michael A. "Biology, Sociobiology, and the Understanding of Human Social Behavior." In *Sociobiology Examined*, edited by Ashley Montagu, 291–310. New York: Oxford University Press, 1980.

Simpson, G. G. *Tempo and Mode in Evolution.* New York: Columbia University Press, 1944.

Smith, C. U. M. *The Problem of Life: An Essay in the Origins of Biological Thought.* London: Macmillan, 1976.

Smith, John Maynard. "Survival through Suicide." Review of *Sociobiology: The New Synthesis*, by Edward O. Wilson. *New Scientist* (28 August 1975): 496–97.

Smith, Sydney. "Studies in the Development of the Rainbow Trout *(Selmo Irideus).*" *Journal of Experimental Biology* 23 (1947): 357–78.

Smocovitis, V. B. "Unifying Biology: The Evolutionary Synthesis and Evolutionary Biology." *Journal of the History of Biology* 25 (1992): 1–65.

———. *Unifying Biology: The Evolutionary Synthesis and Evolutionary Biology.* Princeton: Princeton University Press, 1996.

Snyder, Laurence H. "A Modern Darwin." Review of *Genetics and the Origin of Species*, by Theodosius Dobzhansky. *Ohio Journal of Science* 38 (1938): 47–48.

Spalding, Mark J. Review of *Consilience*, by Edward O. Wilson. *Journal of Environment and Development* 7 (December 1998): 442–49

Spanner, D. C. "On 'Active' Mechanisms in Biochemical Processes." *Physiologia Plantarum* 7 (1954): 475–96.

Stadler, L. J. "The Gene." *Science* 120 (1954): 811–19.

Steadman, David W. Review of *Consilience*, by Edward O. Wilson. *Professional Geographer* 51 (May 1999): 325.

Stebbins, G. Ledyard. "Variation and Evolution in Plants: Progress During the Past Twenty Years." In *Essays in Evolution and Genetics in Honor of Theodosius Dobzhansky*, edited by Max K. Hecht and William C. Steere, 167–92. New York: Appleton-Century-Crofts, 1970.

Stent, Gunther. "Introduction: Waiting for the Paradox." In *Phage and the Origins of Molecular Biology*, edited by John Cairns, Gunther Stent, and James D. Watson, 3–8. Cold Spring Harbor: Cold Spring Harbor Laboratory of Quantitative Biology, 1966.

———. "That Was the Molecular Biology That Was," *Science* 160 (1968): 390–95.

———. "Light and Life: Niels Bohr's Legacy to Contemporary Biology." In *Niels Bohr: Physics and the World*, edited by Herman Feshbach, Tetsuo Matsui, and Alexandra Oleson, 231–44. New York: Harwood Academic Publishers, 1988.

Stern, Kurt G. "Nucleoproteins and Gene Structure." *Yale Journal of Biology and Medicine* 19 (1947): 937–49.

Suley, A. C. E. "Genetics of *Drosophila subobscura.* VIII. Studies on the Mutant Grandchildless." *Journal of Genetics* 51 (1953): 373–405.

Symonds, Neville. "What Is Life? Schrödinger's Influence on Biology." *Quarterly Review of Biology* 61 (1986): 221–26.

———. "Schrödinger and *What Is Life?*" *Nature* 327 (1987): 663–64.

———. "Schrödinger and Delbrück: Their Status in Biology," *Trends in Biochemical Sciences* 13 (1988): 232–34.

Taylor, Charles Alan. "Science as Cultural Practice: A Rhetorical Perspective." *Technical Communication Quarterly* 3 (1994): 67–81.

Teich, Mikulás. "A Single Path to the Double Helix?" *History of Science* 13 (1975): 264–83.

Todorov, Tzvetan. "The Surrender to Nature." Review of *Consilience*, by Edward O. Wilson. Translated by Claire Messud. *New Republic* 218 (27 April 1998): 29–33.

Trefil, James. "Must All That Rises Converge?" Review of *Consilience*, by Edward O. Wilson. *Boston Globe*, 29 March 1998, G3.

Valentine, D. H. "Studies in British Primulas. I. Hybridization between Primrose and Oxlip." *New Phytologist* 46 (1947): 229–53.

van den Berghe, Pierre L. "Bridging the Paradigms: Biology and the Social Sciences." In *Sociobiology and Human Nature*, edited by Michael S. Gregory,

Anita Silvers, and Diane Sutch, 33–52. San Francisco: Jossey-Bass Publishers, 1978.

Vogler, Christopher. *The Writer's Journey: Mythic Structures for Screenwriters and Storytellers.* Studio City, Calif.: Michael Wiese Productions, 1992.

Waddington, C. D. "Some European Contributions to the Prehistory of Molecular Biology." *Nature* 221 (1969): 318–21.

Wald, George. "The Human Condition." In *Sociobiology and Human Nature,* edited by Michael S. Gregory, Anita Silvers, and Diane Sutch, 277–82. San Francisco: Jossey-Bass Publishers, 1978.

Wallace, Bruce. "The Legacies of Theodosius Dobzhansky." In *Genetics of Natural Populations: The Continuing Importance of Theodosius Dobzhansky,* edited by Louis Levine, 39–48. New York: Columbia University Press, 1995.

Washburn, S. L. "Animal Behavior and Social Anthropology." In *Sociobiology and Human Nature,* edited by Michael S. Gregory, Anita Silvers, and Diane Sutch, 53–74. San Francisco: Jossey-Bass Publishers, 1978.

Watson, J. D. "Growing up in the Phage Group." In *Phage and the Origins of Molecular Biology,* edited by John Cairns, Gunther Stent, and James D. Watson, 239–45. Cold Spring Harbor: Cold Spring Harbor Laboratory of Quantitative Biology, 1966.

———. *The Double Helix.* New York: New American Library, 1968.

———. "Succeeding in Science: Some Rules of Thumb." *Science* 261 (1993): 1812–13.

Watson, Martha Soloman. "The Dynamics of Intertextuality: Re-reading the Declaration of Independence." In *Rhetoric and Political Culture in Nineteenth-Century America,* edited by Thomas W. Benson, 91–111. East Lansing: Michigan State University Press, 1997.

Webb, D. A. "What Is Life?" Review of *What Is Life?* by Erwin Schrödinger. N.p., n.d. Reviews of *What Is Life?* Cambridge University Press Archives.

Weiner, Johathan. *Time, Love, Memory: A Great Biologist and His Quest for the Origins of Behavior.* New York: Alfred A. Knopf, 1999.

Welch, G. Rickey. "Of Men, Molecules, and (Ir)reducibility." *BioEssays* 11 (1989): 187–90.

———. "Schrödinger's *What Is Life?* A 50-Year Reflection." *Trends in Biochemical Sciences* 20 (1995): 45–48.

Werner, Michael. Review of *Consilience,* by Edward O. Wilson. *Humanist* 59 (March 1999): 44–45.

Wessels, Linda. "Erwin Schrödinger and the Descriptive Tradition." In *Springs of Scientific Creativity: Essays on Founders of Modern Science,* edited by Rutherford Aris, H. Ted Davis, and Roger H. Stuewer, 254–78. Minneapolis: University of Minnesota Press, 1983.

"What Are We?" Review of *What Is Life?* by Erwin Schrödinger. In *The Lancet* (1945). Reviews of *What Is Life?* Cambridge University Press Archives.

Wilkins, M. H. F. "The Molecular Configuration of Nucleic Acids." In *Les Prix Nobel En 1962,* 126–53. Stockholm: Imprimerie Royale P.A. Norstedt and Söner, 1963.

Wilson, Edward O. "For Sociobiology." *New York Times Review of Books* 22 (11 December 1975): 60–61.

———. *Sociobiology: The New Synthesis.* Cambridge: Belknap Press of Harvard University Press, 1975.

———. "Academic Vigilantism and the Political Significance of Sociobiology." *BioScience* 26 (March 1976): 183, 187–90.

———. "The Attempt to Suppress Human Behavioral Genetics." *JGE: The Journal of General Education* 29, no. 4 (winter 1978): 277–87.

———. Foreword to *The Sociobiology Debate: Readings on Ethical and Scientific Issues*, edited by Arthur L. Caplan, xi–xiv. New York: Harper & Row, 1978.

———. Introduction to *Sociobiology and Human Nature*, edited by Michael S. Gregory, Anita Silvers, and Diane Sutch, 1–12. San Francisco: Jossey-Bass Publishers, 1978.

———. Foreword to *Sociobiology and Behavior*, by David Barash, 2d ed., xi–xii. New York: Elsevier, 1982.

———. *Naturalist.* Washington, D.C.: Island Press, 1994.

———. *Consilience: The Unity of Knowledge.* New York: Alfred A. Knopf, 1998.

———. "Divisive Ideas on 'Unification.'" Interview by Terence Monmaney. *Los Angeles Times*, 9 July 1998, B2.

———. "Even Apes Have a Concept of Morality." Interview by S. Klein and J. Petermann. *Der Spiegel* 46 (1998): 252.

Witkowski, J. A. "Schrödinger's 'What Is Life?': Entropy, Order, and Hereditary Code-scripts." *Trends in Biochemical Sciences* 11 (1986): 266–68.

Wright, Sewall. "Evolution in Mendelian Populations." *Genetics* 16 (1931): 97–159.

———. "The Roles of Mutation, Inbreeding, Crossbreeding and Selection in Evolution." *Proceedings of the Sixth International Congress of Genetics* 1 (1932): 356–66.

———. Review of *Genetics and the Origin of Species*, by Theodosius Dobzhansky. *Journal of Heredity* 33 (1942): 283–84.

Yourgrau, Wolfgang. "Marginal Notes on Schrödinger." In *Biology, History and Natural Philosophy*, edited by Allen D. Breck and Wolfgang Yourgrau, 331–43. New York: Plenum Press, 1972.

Yoxen, Edward. "The History of Molecular Biology." Review of *The Path to the Double Helix* by Robert Olby. *British Journal for the History of Science* 11 (1978): 273–79.

———. "Where Does Schrödinger's 'What Is Life?' Belong in the History of Molecular Biology?" *History of Science* 17 (1979): 17–52.

Index